IRAQ'S SUNNI INSURGENCY

AHMED S. HASHIM

ADELPHI PAPER 402

The International Institute for Strategic Studies

Arundel House | 13–15 Arundel Street | Temple Place | London | WC2R 3DX | UK

ADELPHI PAPER 402

First published February 2009 by **Routledge**
4 Park Square, Milton Park, Abingdon, Oxon, OX14 4RN

for **The International Institute for Strategic Studies**
Arundel House, 13–15 Arundel Street, Temple Place, London, WC2R 3DX, UK
www.iiss.org

Simultaneously published in the USA and Canada by **Routledge**
270 Madison Ave., New York, NY 10016

Routledge is an imprint of Taylor & Francis, an Informa Business

DIRECTOR-GENERAL AND CHIEF EXECUTIVE John Chipman
EDITOR Tim Huxley
MANAGER FOR EDITORIAL SERVICES Ayse Abdullah
ASSISTANT EDITOR Katharine Fletcher
COPY EDITOR Matthew Foley
PRODUCTION John Buck
COVER IMAGE AP Photo

Printed and bound in Great Britain by Bell & Bain Ltd, Thornliebank, Glasgow

British Library Cataloguing in Publication Data
A catalogue record for this book is available from the British Library

Library of Congress Cataloging in Publication Data
A catalogue record for this book is available from the Library of Congress

ISBN 978-0-415-46655-4
ISSN 0567-932X

Contents

GLOSSARY

AM	Army of the Mujahadeen
AMS	Association of Muslim Scholars
AQM	al-Qaeda in Mesopotamia
ASC	al-Anbar Salvation Council
IAI	Islamic Army of Iraq
IED	improvised explosive device
ISG	Iraq Survey Group
ISI	Islamic State of Iraq
JRF	Joint Resistance Front
WMD	weapons of mass destruction

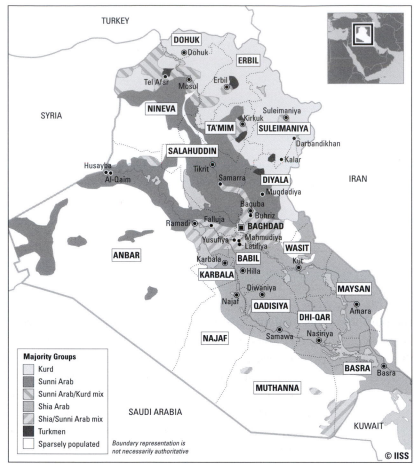

Predominant ethnic–religious groups in Iraq

INTRODUCTION

The purpose of this Adelphi Paper is to trace and analyse the onset and course of the Sunni insurgency in Iraq from 2003 to the end of 2008. The paper provides an overall assessment of the insurgency's causes, evolution, goals, strategies, internal rivalries and structural weaknesses. However, given the dynamic nature of politics in Iraq, it is not clear that any firm and final conclusions or lessons can be drawn at this juncture. The paper addresses the US counter-insurgency campaign only in terms of the role it played in promoting Sunni resistance in the early years, and does not examine whether the US military 'surge' implemented in early 2007 contributed to the diminution of violence in Iraq by late 2008. Nor does it address in detail the roles of Iraq's other ethno-sectarian communities – particularly the Sunni but non-Arab Kurds who make up approximately 18% of the population, and the Shia Arabs who comprise the majority of the population at around 60% – whose attitudes towards the Sunni insurgency have been either ambivalent or hostile. Indeed, the lack of support from or participation of these two large communities is a key factor in ensuring that the insurgency did not emerge as a genuine national-liberation struggle.

In this study, the term 'insurgency' is used to mean conflict between a regular or conventional military and an irregular force that is incapable of facing the former in force-on-force fighting. Irregular forces have to find other ways to match their adversary. In practice, this means that insurgents use guerrilla tactics, defined as hit-and-run attacks on regular forces and attacks on supply lines and small and isolated garrisons. One of the main operational goals of insurgents is to ensure that their forces remain intact. They cannot afford to lose significant forces in combat to more powerful regular forces, which is why they often 'flee' when confronted with regular units. Insurgents sometimes use terrorism – the deliberate use of violence against civilians – as part of their operational repertoire; however, many insurgencies and their leaders have eschewed terrorism as an essential part of their approach. For example, both Mao Zedong and Ernesto 'Che' Guevara argued that the risk that terrorism would alienate the population outweighed its benefits. The third operational means of insurgents is the deliberate targeting of the critical infrastructure of the state and the kidnapping or killing of those manning this infrastructure. Insurgencies are thus wars of the weak against the strong. The strong might be a foreign power that has invaded and occupied another country, or the insurgent group might be mobilised against a local government on account of specific political or socioeconomic grievances.[1] Some countries, including Iraq, have a long history of warfare within their societies.[2]

Iraq: a history of violence

Following the terrorist attacks of 11 September 2001 on the United States, and the subsequent emergence of the Iraqi insurgency, a number of analysts and policy observers in the West began to point to a 'new Arab way of war', an 'Islamic' or even an 'Iraqi' way of war.[3] Even the Islamist strategic thinkers associated with al-Qaeda have written extensively on an Islamic way of war.[4] But this so-called new Arab, Islamic or Iraqi 'way of war' is not new. Irregular warfare has an ancient pedigree in the Middle East; some of these irregular conflicts are well known, while others have not been so well documented.[5]

Mesopotamia, as Iraq was known before the 1920s, has always been a problem for its rulers and for foreign powers. The insurrectionary propensities of Iraq's indigenous inhabitants are rooted in the country's history, which has been punctuated by tribal revolts and insurgencies against any centralising tendencies on the part of whatever government was in charge. One of the more famous historical examples of a frustrated ruler dealing with Iraq's fractious people occurred centuries ago, during the reign of eighth-century Umayyad ruler al-Walid I, when, in a speech well known in Iraq and the wider Arab world, governor of Iraq al-Hajjaj ibn Yusuf al-Thaqafi sternly declared before the obstreperous notables of Kufa: 'I see heads before me that are ripe and ready for the plucking, and I am the one to pluck them, and I see blood glistening between the turbans and the heads.'[6]

Matters did not improve significantly under the Mamelukes, slave soldiers with training in administration and clients of the rising Ottoman Empire.[7] In the eighteenth and early nineteenth centuries, the Mamelukes controlled the few urban areas of Mesopotamia, but most of the countryside was dominated by well-armed tribal confederations, which remained either independent or in a state of near-permanent insurrection. The tribes could not overwhelm the Mamelukes, nor were they capable of sustained collective offensive action, especially against the towns. A balance of weakness was achieved between the centre and the regions.[8] In the 1830s the Ottoman government – now in the throes of a reformist impulse to rectify the weaknesses of the empire – grew alarmed by the Mamelukes' mismanagement and independence and decided to remove them from power. The Ottomans established central control and sought to weaken the power of the tribes through a variety of means, including extensive land reforms, the renovation of the ramshackle administrative and bureaucratic system, co-opting key tribes and the use of military force. Success was uneven, and tribal revolts continued well into the early twentieth century.[9] Despite their reforms, which also included the introduction of secular education, more rational and modern administrative practices and conscription, the Ottomans made no real effort to create a sense of a Mesopotamian state or nation; indeed, the Kurds were left to their own devices in the inhospitable and mountainous north. The Shi'ites were viewed with distaste, a sentiment they fully reciprocated. Furthermore, their integration into the state was well-nigh impossible as the Ottomans viewed them as a front for Shia Iran, the enemy next door. As for the Shi'ites themselves, their clerics – who held a great deal of power – warned their flock not to attend the secular education system that the Ottomans had established.

The British seized Mesopotamia from the Ottomans at the end of the First World War for strategic and economic reasons.[10] There were however profound differences within the British government concerning the future disposition of the large and underdeveloped territory.[11] British prevarication and indecisiveness led disgruntled elements within Mesopotamia to raise the standard of insurrection in 1920: the famous *al-thawra al-Iraqiya al-kubra*, or the Great Iraqi Revolt, whose epicentre was the mid Euphrates region.[12] That insurgency failed, for many reasons. Firstly, there were too many disparate ideological currents, which meant that the rebellion failed to present a united front. Secondly, the tribes failed to create an effective military structure and unified command and control of their forces. Thirdly, the British were able to use divide-and-conquer tactics: many of the tribes of the southern Euphrates region chose to stay out of the fray or even to join the British against their rebellious counterparts. The insurgency was costly for the British, and forced them to take into consideration the political wishes of the inhabitants in the process of forming the new Iraqi state. Nonetheless, numerous insurgencies and revolts continued during the monarchical era, between 1921 and 1958, including a serious insurgency by Shia tribes of the Euphrates in the mid 1930s. At the height of British domination of Iraq, British military forces, principally the Royal Air Force, helped the monarchy keep control by developing and implementing a sophisticated strategy of 'air control', designed to enable the British to maintain government authority over recalcitrant tribes at less cost than using large ground forces.[13]

After the revolution of 1958 ended the monarchical period, a newly republican Iraq faced considerable discontent among the minority Kurdish population in the north. From the early 1960s to the mid 1970s, much of the Iraqi armed forces were geared to fighting a brutal and fruitless campaign against Kurdish guerrillas (the *peshmerga*). The Kurds were defeated in a major offensive after Iran withdrew support and sanctuary as a result of a deal with Iraqi strongman Saddam Hussein in 1975. Benefiting from substantial oil revenues, Saddam implemented a plan between 1975 and 1980 to transform the Iraqi armed forces from a counter-insurgency force into a powerful conventional military.[14] Equipped with high-quality equipment from the Soviet bloc and from some Western nations, the Ba'athist regime launched what it hoped would be a short and victorious war against the disorganised forces of Iran in September 1980. Saddam feared greatly the 'demonstration effect' of Ayatollah Khomeini's revolution on Iraq's own Shia population; he also thought that a quick, successful war against the generally disliked 'Persians' would cement his leadership among the Arabs of the Gulf and within the wider Arab world. The Iraqis were, however, inept warriors, lacking in flair, innovation and decisiveness. Initially, they faced guerrillas and militias in Iran's Khuzestan province who offered stiff resistance. Later, they confronted a more innovative and flexible foe in the revitalised Iranian regular forces and the Islamic Revolutionary Guard Corps. The Iranians decisively defeated the Iraqis in 1982. Iraqi forces withdrew to their own territory and prepared to sue for peace. Khomeini was, however, determined to overthrow Saddam's regime, and moved to continue the conflict. A bloody war ensued, characterised by brutal infantry assaults and massive artillery barrages. In the end, Iraq prevailed due to the superiority of its military resources, support from the international community and sheer exhaustion on the part of Iran.

Saddam erroneously concluded that Iraqi military power had come of age. He was forced to test this assumption in 1991 against vastly superior US and Western forces. Although Iraq failed spectacularly in conventional warfare, its troops succeeded in putting down two poorly coordinated and badly organised insurgencies, by the Kurds in the north and the Shi'ites in the south. For all its enormous conventional arsenal and its supposed experience in regular warfare, the Iraqi Army was put in a strict and stifling straitjacket of political surveillance to prevent it from overthrowing the regime.[15] Moreover, it was not optimised for effective conventional warfare.[16] As the invasion of 2003 loomed, the army set about preparing itself for conventional warfare, but it was a spent force after a decade of sanctions. Meanwhile, the regime did not prepare effectively for irregular warfare against the invading Coalition forces. In any case, it is not clear that, had it been ready, it would have had much success in mobilising Iraqis. Irregular warfare relies on popular legitimacy. Saddam's regime had a great deal of resilience and enormous patrimonial power, reflected in its ability to reward its supporters and punish opponents, but it did not have the popular legitimacy needed to wage a war of national liberation. Some Iraqi units, in particular the Fida'iyin Saddam, did fight desultory partisan campaigns against the advancing Coalition forces, but these were not effective.[17] It was only after the fiasco of the first weeks of occupation, with the breakdown of law and order and massive looting in the cities, that some elements of the Sunni community, particularly officers and officials of the former regime, considered armed opposition to the foreign presence. Mobilised by the belief that the formerly dominant Sunnis were destined to be marginalised, they set about trying to build an irregular or asymmetric response. Thus, for the first time since the creation of the Iraqi state, the dominant community – rather than those at the margins, the Shi'ites and the Kurds – took up arms in an insurgency against foreign occupiers and the Iraqi state. Assessing why and how they did so is the aim of the following chapters.

CHAPTER ONE

Origins, Causes and Composition

This study is about an insurgency by one community within Iraq, the Sunnis. It is, thus, a communal insurgency. It could also be viewed as an 'ethnic' insurgency, as many Sunni insurgents see themselves as fighting the empowerment of both non-Arab Kurds and Shi'ite Arabs whose Arabness they regard as compromised by their religious affiliation.[1] It is not a separatist insurgency, since no insurgent organisation or Sunni political movement has called for the creation of an independent or autonomous 'Sunnistan'. In fact, many Sunni insurgent groups objected when radical Islamists under the aegis of al-Qaeda in Mesopotamia (AQM) declared the establishment of an 'Islamic State of Iraq' (ISI) – effectively, an insurgent group of the same name – in mid October 2006, believing this to be a prelude to secession from Iraq. The Sunni insurgency did not emerge from a politically and socioeconomically marginalised ethnic or communal group. Rather, it is that rare phenomenon, an insurgency by a hitherto dominant group seeking to restore its former position of power: a restorationist or 'reactionary' insurgency.[2] In short, the insurgency began with a reaction by a group dismayed by the *loss*, rather than by a simple lack, of power, privilege and reward.

Origins and causes

An identifiable set of factors prompted the onset of the Sunni insurgency. Conceptually, we can categorise origins and causes in two distinct ways. Firstly, they can be classified along a continuum, from the basic to the more complex. There are some straightforward reasons for the insurgency, such as resistance on the grounds of opposition to foreign occupation. Other reasons are more complex, and are founded on ideological principles or justifications; in the early years of the occupation, the insurgents were unable to formulate and present their beliefs and goals, but since 2006, insurgent groups have gone to great lengths to articulate their aims, often on the Internet or in extensive interviews with Iraqi, regional and international media. Secondly, we can distinguish between causes that are material in nature – 'I am fighting because I lost my job/salary or because you (the Americans) dissolved the army' – and those that stem from complex identity concerns – 'I am fighting to defend the Sunni community from marginalisation and the takeover of "our country" by the Shi'ites and the Iranians.'

The most fundamental cause of the Sunni insurgency was resistance to foreign occupation. One is reminded of a (possibly apocryphal) conversation between a Union officer and a captured Confederate at the end of the American Civil War. The Union officer asks the Confederate why he is fighting. He answers 'because you are down here' (i.e., in the South). Numerous statements by Iraqis attest to the fact that they are fighting what they perceive to be an occupation. For example, in an interview in 2005, a detained Iraqi schoolteacher in the town of Tel Afar in Nineva province said that he was fighting occupation because it was illegal and unjust; 'what would you do', he asked his interviewer, 'if I had invaded your country?'[3] In the early stages of the insurgency, a tribal leader told a Western journalist that 'the entire Iraqi people is a time bomb that will blow up in the Americans' face if they don't end their occupation. We refuse to deal with the occupation.'[4] When asked why he was fighting, a member of one of the largest and most prominent insurgent groups, the Islamic Army of Iraq (IAI), responded: 'I want you to ask this question to the US forces, not to me. They came from the other side of the world and crossed the ocean to occupy my country.'[5] The insurgent, Abu Ayoob, went on to say that the Coalition had 'no credibility', concluding, like many Iraqis, that it had come to Iraq 'to destroy Islam, steal oil, save the east front of Israel, control the Middle East and establish bases near Iran and Russia', rather than to promote democracy.

Another important factor prompting Sunni resistance was the early American policy of either ignoring the Sunnis or of seeing them collectively as the 'enemy' because they had been the mainstay of the deposed regime. American indifference or hostility towards the Sunnis was deliberate in the early days of the occupation, and led to their marginalisation in accordance with the Bush administration's strategy of depending on the more 'reliable' Shia Arabs and Kurds, who together constituted a majority of the population. The Sunnis realised this from the very beginning, and were determined not to be ignored. If this meant taking up arms, then so be it, and they had a formidable capacity for resistance. Once the Sunnis went into rebellion, the initial heavy-handedness of American forces in the field exacerbated and perpetuated their resistance.[6] The culture clashes, mutual misunderstandings and rough treatment of civilians proved disastrous. Explaining why he had joined the insurgency, Abu Ayoob said: 'when the occupation forces entered Baghdad, they killed my brother in front of my eyes. He was wounded and bleeding but the occupation forces didn't allow me to save him. When I tried to save him they began shooting at me and after a few minutes my brother died. After that I swore to fight them to the death.'[7]

The origins of the Sunni insurgency cannot be understood without reference to the evolution of the Iraqi polity from Ottoman times to the present. The Ottomans were responsible for the emergence of the Sunni Arabs as the dominant political community in Mesopotamia from the mid nineteenth century. At the same time, they distrusted the Shia Arabs, who they saw as a fifth column for the Persian Safavids, a Shia dynasty with whom the Ottomans – the pre-eminent Sunni power – were fighting for hegemony over the Islamic *umma* (community) and control of strategically located Mesopotamia.[8] The British further institutionalised and perpetuated Sunni Arab domination over Iraq, though Shia Arabs formed a majority of the population due to the slow but steady conversions to Shiism among the Bedouin Arabs of the south from the late eighteenth century onwards.[9] Over time, the Shia were marginalised through their removal from the centres of political and economic power: Sunnis controlled the educational and cultural evolution of Iraq to such an extent that Iraqi nationalism came to be equated with closer links with and integration into the wider, largely Sunni, Arab world. For the Sunni Arabs this was a way of compensating for their demographic weakness and enhancing their political stature. For the Shia Arabs, Sunni-dominated Iraqi Arab nationalism was a further vehicle for their reduction to the effective status of a minority.[10] In short, Sunni Arabs viewed Iraq as 'their' country, one that they had formed and shaped. Shia Arabs were viewed as Iraqis, but they could be neither the rulers nor the shapers of Iraq's identity. There was and is among Sunnis considerable fear about the 'Shia-isation' of Iraq. Adnan al-Dulaimi, head of the Sunni *awqaf* (charitable endowment) in Iraq, gave an interview to a Saudi newspaper in 2005 in which he argued that the Sunnis faced a major conspiracy against their identity at the hands of the occupation and the 'shu'ubis' (Iranians), who supposedly have a racist hatred of Arabs:

> These acts are nothing but the work of the new Shu'ubism, which is trying to abolish the Arab character of Iraq and make it part of the world that is in the hands of a colonialist hegemony … I would like to conclude [the interview] by appealing to all Arab countries to be on their guard about the conspiracies being hatched against this region, that what is taking place in Iraq is nothing but a prelude to the shape of things that might eventually take place in their own territories, and that the fire that might eventually destroy Iraq might also engulf the whole region.[11]

The post-occupation marginalisation of the hitherto dominant Sunni Arab community paralleled the steady political empowerment of the formerly marginalised majority Shia community.[12] The Shia takeover of the Iraqi state, particularly the military and security services, has played a major role in the radicalisation of Sunnis, recruitment into the ranks of the insurgency and the rise of myriad neighbourhood self-defence militias to protect Sunni civilians from Shia militias and death squads and Shia members of the official Iraqi security forces.[13] There was little that the US could do to thwart the Shia takeover of the security forces without antagonising the major Shia parties, many of whose members declared that Shia control of the security forces in the post-Saddam era was not open to negotiation.

By the end of 2006, for many Sunni groups, the insurgency was no longer merely a struggle against the Coalition presence; rather, it was slowly being transformed into what was viewed as a defence of the Sunni community against a rising Shia tide. The Salafist[14] insurgent umbrella group the Mujahadeen Shura Council declared in July 2006 that it was going to create new combat brigades tasked with the goal of 'eradicating' the Jaish al-Mahdi (Mahdi Army) Shia militia led by cleric Moqtada al-Sadr. The new brigades would come under the command of the Umar Brigade, a militia established in 2005 to eliminate members of the Shia Badr Brigade. Other Sunnis began to talk of creating their own militias or of asking insurgent groups to provide security for their community. Still others began to reflect that there might be merit to the presence of US forces – a hitherto outlandish idea for Sunnis – as the American presence acted as a brake on Shia expansion. In late November 2006, AQM led an assault on the largely Shia Sadr City in Baghdad in revenge for attacks on Sunni neighbourhoods. Notwithstanding serious rifts within the Sunni insurgency itself, five Sunni groups were involved in launching the counter-assault against the Shia, including groups not affiliated with AQM, such as the IAI and the Just Retribution Brigades. In late November 2006, AQM creation the ISI issued a statement justifying its attacks on Shi'ites and the Coalition in language that even its Sunni opponents had no difficulty in supporting:

> Since the beginning of the Crusader invasion of Iraq, our country, the best of the Sunni mujahadeen have been busy pushing the attacking enemy out of their country. The arrows of the mujahadeen focused on the occupying Crusader enemy until the black hatred of the Shia in Iraq for the Sunnis became obvious to all eyes. This sect, alongside the invading Crusader forces, became the tip of the spear in the fight against the mujahadeen.[15]

The Shi'ites have been attacked for 'treacherously' allying themselves with Christians and Jews. The ISI has also said that the conflict with the Shi'ites was a decisive battle 'between Islam and unbelief', crucial not only for the survival of Iraq's Sunnis but for the 'whole Islamic world'. The Islamic world could not allow Baghdad, the home of the caliphate, to become a 'Shia city where the mosques of monotheism are destroyed and burned and paganism is preached'.[16] Despite their profound differences on political, ideological and methodological issues, the Sunni insurgents seemed to be united about the nature of the dangers facing the Sunni community in Baghdad and elsewhere.[17]

Ultimately, the insurgency has both material and identity-related causes. Sunnis have been fighting to prevent the material takeover of 'their' state by Shi'ites, whom they believe to be working to irrevocably change the identity of the country.

Composition

Far from being one monolithic organisation, the Sunni insurgency has consisted of different movements. The dividing lines are blurred, shifting alliances are common and many insurgent groups have cooperated or worked with others that are ideologically different or driven by dissimilar goals. For the sake of simplicity, however, the insurgency can be divided into five distinct elements.

Ba'athists and their affiliates

The first group consists of Ba'athists and associated elements of the former regime. This category includes former regime officials, officers of the previous Iraqi Army, members of the former regime's two 'praetorian' guards (the Special Republican Guards and the larger Republican Guards Corps), security and intelligence officers and paramilitary organisations such as the Fida'iyin Saddam.

Following the collapse of Saddam's regime in 2003, the Ba'ath Party is no longer a unified movement. It has been riven by internal strife within the remnants of its senior ranks, most of whom are ensconced in the Syrian capital, Damascus. Some of the party leadership has coalesced around Izzat Ibrahim al-Duri, a former senior regime official. Al-Duri, who had a reputation as the most religiously minded member of the former regime, has sought to articulate the policies and beliefs of the party since its fall from power. When Saddam was executed at the end of 2006, a large number of former Iraqi Ba'ath Party functionaries gathered in Syria to mourn the 'martyrdom' of the former leader and to declare al-Duri his rightful and legitimate successor as president of Iraq. He has not gone unchallenged. Another group has formed around Mohammad Yunis Ahmed al-Muali, a former high-ranking Ba'ath Party intelligence official. Ahmed was responsible for party activities in Salahuddin, Ta'mim and Suleimaniya governorates during Saddam's regime. He has emerged as a key financier and organiser of his splinter group, known as the New Ba'ath Party.

In addition to internal divisions, the weaknesses of the Ba'ath Party stem from several other sources. Firstly, it has lost power and its original ideology is no longer of much relevance. Secondly, the exiles' host government, Syria, has sought to foment factionalism in order to maintain better control over what remains of the party leadership. Syria does not wish to give the US a pretext to take action against it, as it might have done had Syria promoted a unified and effective Ba'ath Party as a credible opposition. Furthermore, Syria is allied with Iran, and has no wish to alienate Tehran by promoting a Sunni-dominated party under whose rule Iraq launched the war against Iran in the 1980s.

During its time in power, the Ba'ath Party created a number of armed militias and paramilitary forces. Their level of military effectiveness varied considerably; for the most part, these groups were designed to shore up the regime's waning popularity.[18] Once the party fell from power in 2003, it set about creating groups dedicated to waging irregular war. There are reported to be several insurgent groups affiliated with the Ba'ath Party. Immediately following the fall of the regime, a group of former Ba'athist officials, intelligence operatives and ex-officers formed a shadowy organisation called al-Awda (The Return), which became one of the first insurgent groups to carry out attacks on Coalition forces. The Fida'iyin Saddam, which re-emerged from four years of relative obscurity in late 2006, can be seen as a military wing of the Ba'athists. Established in 1995, the group was comprised of barely literate young men from regions loyal to the regime in Sunni-majority Anbar province in central Iraq. Since the demise of the Ba'ath Party, Fida'iyin Saddam members have adopted a fusion of Ba'athist nationalist rhetoric and Islamism. Jaish Muhammad (Muhammad's Army) can also be considered a military wing of the Ba'athists. It was established shortly after the collapse of the regime, staffed by former high-ranking officials and officers. It had roughly 1,000 members, but suffered heavy casualties at the hands of Coalition forces during the two battles for control of the city of Falluja in 2004. It has become heavily influenced by Islamist beliefs, and its links with the Ba'ath Party have become looser and less formal.

Nationalist-Islamists and their affiliates

The second group consists of nationalist-Islamist organisations, made up of Iraqis who oppose the Coalition presence on patriotic, nationalist and Islamic lines. These organisations range from those that believe that they have a legitimate right to resist occupation and do not feel it is necessary to develop a more complex ideological formulation for their resistance, to actors who have articulated solid nationalist and Islamist principles to justify their resistance activities. This group includes intellectuals and members of the middle class, as well as former Ba'athists who became disillusioned with the party over the course of its last decade in power, and who believe that its pan-Arab nationalist ideas have no relevance to Iraq's current situation. This disillusion with the Ba'athist ideological edifice was sealed by the regime's collapse in 2003 and Saddam's humiliating capture later that year. There has been considerable cooperation and overlap between the Ba'athists and nationalist-Islamist groups.

Iraqi Salafists and their affiliates

The third group consists of Iraqi Salafist Islamist organisations. Salafism promotes a philosophy of *tawhid*, the absolute unity of God, and excoriates any system that adopts man-made rules and laws and fails to rule in accordance with sharia (Islamic law). Over the past decade, Iraq has witnessed a growth in Islamist sentiment, for reasons discussed in the next chapter. The fall of the regime removed the constraints on the overt emergence of Salafist Islamist movements within the country. Two other factors reinforced the rise of Salafism. Firstly, the institutionalisation of the Coalition occupation and the promotion of Western political ideas in Iraq contributed to the emergence of an opposition based on Islamist beliefs and values. Secondly, with the end of the Sunni domination of Iraq and the political and religious empowerment of the Shia community, many Sunnis felt that opposition to these disruptive events could only come from within the framework of Salafist Islam. The largest and most important of the Salafist groups are Ansar al-Sunna, the 1920 Revolution Brigades, Jaish al-Rashidin and the IAI.

Ansar al-Sunna is one of the most active of the local Islamist insurgent organisations. It was formed in November 2003 under Abu Abdullah Hasan Bin Mahmud as the Sunni Arab counterpart to the Kurdish Islamist group Ansar al-Islam. Most of its leadership initially gained operational experience with the Kurdish group. Its most notable early operations included the bombing of the Patriotic Union of Kurdistan and Kurdistan Democratic Party headquarters in Erbil in the north in February 2004, in which 100 people were killed. Ansar al-Sunna is made up of a number of *kata'ib* (brigades), the most prominent of which are Abu Hanifa Nu'man, Asad al-Islam and Sa'ad Bin Abi Waqqas. The towns of Ramadi, Latifiya, Yusufiya and Mahmudiya and the province of Diyala, where much of the fighting in 2007 took place, are its major operational centres.

The 1920 Revolution Brigades comprises nationalists – many of them former regime loyalists who have abandoned Ba'athism – and Islamists. It was founded in July 2003, in the opening stages of the insurgency. It is organised into several combat brigades, and has maintained a high operational tempo. It is also closely linked to the mainstream Sunni political parties and organisations involved in the political process in Baghdad. In March 2007, the 1920 Revolution Brigades split into two groups. The original 1920 Revolution Brigades maintained its Salafist orientation, while the other group to emerge from the split, which called itself the Islamic Resistance Movement: Hamas-Iraq, stated an intention to create parallel political and military wings to conduct both 'armed jihad and political action to achieve the goals of the resistance'.[19]

According to its commander, Jaish al-Rashidin was formed in the early days of the insurgency. It is very similar in worldview and structure to the 1920 Revolution Brigades.[20] It comprises several combat battalions and brigades and has units in many provinces.

The largest Salafist organisation is the IAI, which brings together former regime loyalists, nationalists and local and foreign Islamists. Its unwieldy size and the presence within it of myriad groups with different ideologies have led to factionalism and the splitting off of some elements. One splinter, the Al-Fatihin Army, was established in February 2006. The Al-Fatihin

Army targets the Coalition and Shia militia. It has cooperated with other insurgent groups, including the Jaish al-Rashidin, Jaish Muhammad, the Supreme Command of the Armed Forces (composed of former officers from the previous Iraqi Army) and the Joint Leadership of the Mujahadeen. In the latter half of 2007, another group, Jaish al-Furqan, split from IAI over differences concerning the direction and modus operandi of the parent unit.[21]

Iraqi tribes

The definition of a tribe and the characteristics of such a group have been the subject of considerable debate in anthropology and sociology.[22] Briefly, a tribe is a group of people who share, or feel that they share, common kinship ties and a common history. Notwithstanding some Marxist and Western modernisation theories that tribes and their norms were atavistic manifestations of pre-modern societies destined to disappear, tribes have remained intricately woven into the fabric of society in many countries in the Middle East, including Iraq.

As we have seen, the relationship between the tribes and the central authorities in Iraq has ranged from mutual cooperation to ambivalence and outright hostility. Whenever the state sought to increase its power and authority, the tribes would resist and there would be conflict. Equally, when state power and capacity contracted due to financial and admin-istrative weakness and lack of effective coercive capacity, the power of the tribes would expand. If state formation and state-building in Iraq had followed a trajectory of ever-increasing centralised power, the strength of the tribes would probably have been broken. But the power of the state in Iraq has waxed and waned, strong at some times, weak at others. And while the Ottomans succeeded in destroying several tribal confederations, the British revived the tribes as an alternative power centre to the obstrep-erous urban nationalists and the monarchy.

Following the 1958 revolution, the first republican regime generally ignored the tribes or tried to restrict their power. But under subsequent governments, particularly that of the Arif brothers in the 1960s, tribal ties were woven tightly into the power structure of the ruling military elite. The Ba'athist regime came to power in 1968 with the belief that the tribes represented a reactionary force in Iraqi society and constituted an obsta-cle to modernisation and progress.[23] At the same time, however, Saddam himself was very aware of his own tribal origins and of the important role that kinship ties and tribal values could play in support of the state and his own power base. From the very early days of the Ba'athist regime the then-largely ceremonial Republican Guard was filled with recruits from Saddam's tribe in Tikrit, the Albu Nasir. Later, Saddam relied on the tribes for the manpower he needed for the massive expansion of the Iraqi military during the Iran–Iraq War. In return, the tribal sheikhs demanded resources for themselves and their tribes. By the 1990s, as sanctions ate into the capacity of the state and reduced its ability to maintain control over society, Saddam began to rely on loyal tribes to keep himself in power. Furthermore, as Iraq's society began to collapse under the weight of international sanctions, many urban people returned to their tribes, which acted as a kind of social-security umbrella. This enhanced the prestige of the tribes, but it also put enormous economic pressure on them. The tribes began to play an increasingly significant role in Iraq's shadow or illicit economy.

On the eve of the 2003 invasion, the tribal sheikhs declared their loyalty to Saddam and the regime and promised material support. But the sheikhs were also pragmatic and prudent, and knew that the occupation presented opportunities as well as dangers to the tribes. The dangers stemmed from several sources: the collapse of security and stability, the United States' clear lack of understanding of the importance of the tribe in Iraq's fragile society, and the rise of the Sunni insurgency. The opportunities lay in the potential role the tribes could play in security, stability and governance at local levels, with the Americans working with and empowering them. Many tribal sheikhs ensured that governance and the provision of basic services were maintained in their areas following the collapse of the regime. There was, however, no official strategy to incorporate them into the new system of patronage that Iraq's occupiers were setting up. Furthermore, the US did not appreciate or understand the role tribes and their sheikhs could play in national politics at the centre. All of this struck at the tribes' sense of identity and self-worth. As the American presence expanded into the tribal domains, the sheikhs and young men of the tribes began to chafe under the dishonour of occupation, and several tribes joined the insurgency. The Americans responded in a heavy-handed manner, affronting tribal notions of self-esteem, honour and manhood.

Despite these tensions, there was nothing in principle precluding a pragmatic accommodation between the tribes and the occupying forces, and by 2005 the tribes had shown that they were willing to come to terms. The impetus for the emergence of bargaining between tribe and occupier was provided by a force that seemed more menacing to the tribes' well-being than the Coalition. This force was AQM, whose actions threatened the spectrum of tribal interests and identity. The radical Islamists encroached on tribal economic interests, extracting protection money from merchants and travellers, imposing taxes and shutting down barber shops and coffee houses. They also imposed strict doctrinal principles on tribal culture and affronted tribal norms and conceptions of honour and dignity by publicly flogging or beheading tribesmen who strayed from their doctrinal path. As one US journalist put it: 'We've learned an important lesson in Anbar province: the Islamic-extremist message is a loser. Most Muslims do not want to live without music, television and, especially, tobacco. They don't want their daughters forcibly married to jihadis or their sons shrouded in explosive vests.'[24] When relations between the tribes and AQM became violent in 2006–07, and the tribes began to set up the collective fronts against AQM that were to become known as 'Awakening Councils' (sahwa), they turned to the Americans for help.

Transnational Salafi jihadists associated with al-Qaeda
The fifth group consists of AQM and its affiliates. These groups subscribe to a rigid and inflexible form of Salafist Islam. Unlike the more 'moderate' local Salafists, this group does not hesitate to pronounce *takfir* against (that is, excommunicate, or declare to be non-Muslim) Muslims who do not subscribe to its ideology. This group has been profoundly hostile to the Shi'ites, who they refer to as *rafidah* ('rejectionists' of Islam).

In the early stages of the insurgency, most of the Salafi jihadists (or *takfiris*, as they were often known) were non-Iraqis led by the Jordanian Abu Musab al-Zarqawi and his organisation, Tawhid wa al-Jihad (Monotheism and Holy War). They were not linked with or subordinate to Osama bin Laden's al-Qaeda organisation. Al-Zarqawi conducted a large number of operations, including the high-profile kidnapping and execution of foreign hostages – most notoriously the videotaped killing of the young American Nicholas Berg – as well as attacks on the Shia-dominated security forces and Shia civilians. In 2005, al-Zarqawi brought his group under the loose control of bin Laden's movement. The merger served both organisations' interests, providing al-Qaeda with a ready-made base from which to strike the US in Iraq, and giving al-Zarqawi prestige, recruits and financial and logistical support. The movement became AQM.

AQM maintained a high operational tempo and conducted a wide range of operations across the country. It also sought to ensure that it emerged as the dominant Sunni insurgent group. To that end, in January 2006 it established an umbrella organisation, the Mujahadeen Shura Council. In October 2006, it set up the ISI, and reiterated its call to other insurgent groups to give their allegiance. The jockeying for power within the insurgency and AQM's tactic of inflicting mass civilian casualties allegedly caused concern within the leadership of al-Qaeda itself, and eventually brought AQM into conflict with other Sunni insurgent groups and with the Sunni tribes of Anbar province.

The presence of a plethora of insurgent groups in Iraq has had adverse implications for the articulation of a unified ideology and set of goals. It is not clear what the myriad groups had in common with one another at the beginning of the insurgency beyond a desire to rid the country of the foreign presence, fight the rise of the Shia majority and ensure a return to Sunni hegemony. The wide variety of ideological currents within the insurgency was a reflection of deep divisions and fissures within the Sunni community, divisions which had barely been kept in check during the last decade of Saddam's rule. The following chapter will explore these currents.

CHAPTER TWO

Ideology

An effective insurgency requires ideological unity and coherence. As we have seen, these qualities have been lacking in the case of the Sunni insurgency in Iraq. Indeed, no insurgency in recent memory has been as factionalised and divided ideologically.[1] The ousted Ba'ath Party lacked support and was unable to mobilise Sunnis, leaving a vacuum that was filled by the nationalists, Islamists and tribes. The emergence of transnational Islamists under the banner of al-Qaeda added another layer of complexity to the insurgency, while at the same time resolving the ideological divisions within it into two major groups: those that were fighting for purely Iraqi goals – removing the Coalition (later modified to securing the support of the Coalition against the Shi'ites) and ensuring the return of Sunni dominance (later modified to ensuring that the Sunnis received a share of the spoils) – and those ranged on the side of AQM, who were fighting to implement a theocracy and seeking to promote Islamist goals beyond Iraq.[2]

Ba'athists and their affiliates

When it was founded in the 1940s, the Ba'ath Party adopted the slogan 'Unity, Freedom and Socialism'. These were understood as integrated elements, none of which could be attained without the other two. Of the three, however, Arab unity was considered the most important goal, being the primary element of Arab rebirth. The doctrine of a single, indivisible Arab nation was central to Ba'athist ideology; 'mere' statehood (i.e., the existence of several independent Arab states) was regarded as parochial, reactionary and a factor contributing to Arab weakness and susceptibility to domination by outside powers. Socialism was not an end in itself, but a means for achieving the higher ends of freedom, unity and social and economic justice. Freedom meant national freedom from foreign control, poverty and hunger, not the freedom of the individual to do as he wished. On the tricky problem of Islam, Michel Aflaq, the key founder of Ba'athism, wrote that Islam was to be glorified as a religion and a worthy cultural and historical heritage, but that it could not be part of politics. In foreign policy, the party advocated non-alignment and neutrality.[3]

This was the theory. The reality of Ba'athist rule in Iraq under Saddam was more complex. Some Ba'athist dogma was initially followed, such as socialism and the call for Arab unity through the subversion of existing states. However, by the mid 1970s, Saddam's regime had veered away from socialism and had begun to accept the existence of several independent Arab states, and to call for greater Arab cooperation and coordination in the face of threats from outside powers, particularly Iran. The regime favoured the control of religion by the state and suppressed manifestations of political Islam. The fundamentals of Ba'athist principles governing the relationship between the state and the mosque were defined after major Shia disturbances in 1977, when Saddam himself, ostensibly then only second-in-command in the regime, expounded the Ba'athist standpoint in a booklet entitled *Nashra fi al-din wa al-Turath* ('A Glance at Religion and Heritage'). In this, he put forward three guiding tenets. Firstly, while the Ba'ath Party was not opposed to religion, it was a temporal party and would remain so. Secondly, the politicisation of religion was impermissible: religion should remain the private affair of the individual. Thirdly, the upkeep of religious sites apart, the state should not interfere in religious affairs. Involvement in religion would compel the state to take sides in religious disputes, which could lead to the revival of centrifugal forces in Iraqi society capable of tearing the country apart.[4]

The First Gulf War, in which most Arab states sided with the US-led international coalition, sounded the death knell for the key Ba'athist dogma of Arab unity and coordination in the face of foreign powers. Even Iraqis who did not support the regime's assault on Kuwait were stunned by the Arab 'betrayal' of Iraq. The years of sanctions during the 1990s and the massive corruption and patronage instituted by the regime ended any ideological illusions; Iraqis focused merely on survival, and began to turn in large numbers to religion. The ignominious collapse of the regime in April 2003 was a signal to the remnants of the Ba'athist hierarchy that their ideology was obsolete and change essential. There is now a general recognition that the pre-invasion system cannot simply be restored. This lies behind Izzat Ibrahim al-Duri's promotion of the idea that the Ba'ath Party 'has undergone an internal shake-up, restructuring its base and leadership on struggle-oriented, faith-based patriotic and nationalist principles. It now has a revolutionary, struggle-oriented identity and has shaken off the dust of the past'.[5] Under al-Duri, it appears that the party has made religious faith an element of its new ideological stance. It has also rejected the current political process in Iraq as a creation of the occupiers, arguing that the entire Iraqi people must be mobilised to eject the occupiers. Following the removal of foreign occupation forces, the party intends to restore Ba'athist rule. However, even a reformed Ba'athism cannot create a major basis of mobilisation for the insurgency or for a 'post-liberation' political entity. The Ba'athists have thus drawn closer to those nationalist-Islamist elements of the insurgency that have not shown irrevocable hostility to the former regime and its ideology.

The Ba'athists and those close to them within the nationalist-Islamist strand are torn between their desire to appeal to the rest of the country's ethno-sectarian groups, in order to present a united front in opposition to the foreign presence in their country, and their fear of domination by the Shia majority. Just as it did among other Sunni groups, a chauvinistic Sunni Arab nationalist streak crept back into the discourse of some Ba'athist and nationalist-Islamist groups in parallel with the Shia takeover of the political process and the security services, and the increased influence of Iran.[6] Many mainstream Ba'athist and nationalist insurgents now view the Shi'ites and Iran as a combined threat equal to or greater than the American presence. The new emphasis on Islam among some Ba'athists stands in marked contrast to past suspicions of political Islam, and reflects the growth of religious sentiment among party members, particularly during the years of occupation. Al-Duri recognises this, and applauds the fact that the Islamist element of the insurgency has played a critical role in obstructing the Coalition. At the same time, however, while they have drawn close to the nationalist-Islamists and even to some Iraqi Salafists, the Ba'athists are opposed to some of AQM's methods – particularly the mass targeting of civilians, which has contributed significantly to the rise in communal violence – and to the organisation's goal of creating a Sunni theocracy.

Nationalist-Islamists and their affiliates

This group has strongly opposed the Coalition presence, but not within the framework of what it sees as a discredited Ba'athist ideology, or within that of militant political Islam. These groups chose instead to fuse a nationalist and an Islamist narrative into a rhetoric of resistance that sought to transcend Iraq's ethno-sectarian divisions and unite all Iraqis in a national *muqawama,* or war of resistance/national liberation. Many of the early insurgent leaders in this group were unemployed former army officers, who had become more pious during the 1990s as international sanctions undermined living standards and destroyed their hope in the future. When the insurgency broke out in 2003, many of its commanders understood it in terms of legitimate resistance to foreign occupation. In an interview with an Arabic newspaper in late 2003, one insurgent leader, a former officer, explained his motivations, which were largely nationalistic:

> The Iraqi land is currently under occupation. Therefore, it is our duty to sacrifice our blood for it. The occupation harms our dignity and people. Many of our kinsfolk have been killed. So, what else is there to live for? … The Americans are in a predicament, and we will continue to beat their heads until they admit the fact that they are greedy occupiers who implement Israel's policies … The struggle for the homeland has been decreed upon us. Our fathers and grandfathers fought against British colonialists and we now fight against the Americans. No occupier will remain in our homeland, no matter how long it takes.[7]

Patrick Graham, a Canadian journalist who spent a year with Iraqi insurgents in the early years of the insurgency, quotes another insurgent leader as saying:

> When we see the US soldiers in our cities with guns, it is a challenge to us. America wants to show its power, to be a cowboy … Bush wants to win the next election – that is why he is lying to the American people saying that the resistance is Al Qaeda … I don't know a lot about political relations in the world, but if you look at history – Vietnam, Iraq itself, Egypt, and Algeria – countries always rebel against occupation … The world must know that this is an honorable resistance and has nothing to do with the old regime. Even if Saddam Hussein dies we will continue to fight to throw out the American forces. We take our power from our history, not from one person.[8]

For many insurgents within the nationalist-Islamist strand, the capture of Saddam in mid December 2003 came as a shock, even if they had not viewed him as important to the resistance. For others, it meant liberation of the insurgency from the tyrant's shadow. One insurgent leader in Baghdad told a journalist that 'Saddam was nothing to our group. We do not like Saddam. But we cannot sit doing nothing while our country is under occupation by foreign forces destroying our culture'.[9] This commander, a 54-year-old university professor, had fought in the war against Iran, in which his brother had been killed. He did not have kind words for the Salafist extremists of AQM either: 'I believe in God and my religion is very important to me, but I am not a religious extremist. These people [AQM] are very dangerous. They want to kill all westerners in Iraq, even civilians, and Iraqis who cooperate with the Americans.'[10]

The nationalist-Islamists, like the Ba'athists, attack the Shi'ites as a fifth column for Iran and as betrayers of Arab nationalism. But neither attacks the Shi'ites on religious grounds, as both local and transnational Salafists do. In 2006, a nationalist insurgent expressed his views about the future trajectory of the country to an Iraqi journalist, arguing that the violence in Iraq was a war between the country's various communities:

> Look, a full-scale civil war will break out in the next few months. The Kurds only care about their independence. We the Sunnis will be crushed – the Shi'a have more fighters and they are better organized … they are supported by the Iranians. We don't have leadership and no one is more responsible for our disarray than Zarqawi, may God curse him.[11]

As for Iran:

> When we feel that an American attack on Iran is imminent, I myself will shoot anyone who attacks the Americans and all the mujahideen will join the US army against the Iranians. Most of my fellow mujahideen are not fighting the Americans at the moment, they are too busy killing the Shia, and this is only going to create hatred.[12]

Iraqi Salafists

The ideological origins and foundations of the Iraqi Salafist current are unclear. But any investigation should certainly begin with reference to the emergence of religion in Iraqi sociopolitical life, particularly during the last decade of the Ba'athist regime, whose efforts to suppress the politicisation of religion failed. While as an ideology and a tool for mobilisation religion was not as well organised among the Sunnis as it was among the

Shi'ites, the Ba'athist regime did not succeed in eradicating Sunni Islamism as a political phenomenon during its long period in power.

Several major events contributed to the revival of religion among Sunnis. The first was the impact of the Iranian Revolution and the Iran–Iraq War. The Iranian Revolution described itself in ecumenical terms as an 'Islamic revolution', rather than as an exclusively Shia one, partly in recognition of the fact that it would have a greater 'demonstration effect' if it appealed to the wider Islamic world, particularly the Sunni Arab countries. Naturally, though, the revolution's largest impact was on the Shia populations in Arab countries. The Ba'athist regime responded to Iran's revolutionary theocratic message by strengthening its Arab nationalist discourse. The regime could not very well attack Shiism itself as the majority of the country's population was Shia, as was the vast majority of the lower ranks of the Iraqi military. The Shi'ites proved their loyalty to the Iraqi nation by fighting tenaciously for eight bloody years against their Shia brethren in Iran.

Secondly, the humiliating military defeat of 1991, a massive psychological blow, caused many Iraqis to question official dogma. The social disintegration that followed under the sanctions regime saw rising impoverishment and crime, the destruction of values, the collapse of the secular state education system and the departure of large swathes of the professional classes. The collapse of the education system was particularly devastating. In 1990, the school enrolment rate was 56%; by 1994 it had fallen to 26% as free education was withdrawn and parents sent their children out to beg or look for work. The adult literacy rate, 89% in 1985, fell to 59% in 1995. Young people either did not start higher education or dropped out mid-course because they needed to work; adults with higher degrees were forced to make a living by non-professional work.[13] As one foreign journalist put it in 1995: 'in the Shorje market in downtown Baghdad, it is not uncommon to find academics, accountants, and engineers selling powdered milk or soda pop'.[14] This decaying social and economic environment led many Iraqis to turn to religion for succour and spiritual comfort.

Thirdly, Saddam himself initiated a 'return' to religion, known as 'al hamla al-imaniyya' (return to faith),[15] despite his earlier trenchant opposition to the mixing of religion and politics and arguments against the institutionalisation of religion in state and society. Given the dire circumstances in which the country found itself in the 1990s, the regime felt it necessary to call upon Islam in order to ensure its survival – Ba'athist ideology was discredited in the aftermath of the disaster in Kuwait, and a new ideological underpinning had to be found. Under the 'Return to Faith' campaign, the government promoted mandatory Koranic studies in schools and built training centres for imams, including Saddam College and the Saddam University of Islamic Studies. Radio stations regularly broadcast Koranic lessons, and alcohol was banned in restaurants. The steady rise in religious sentiment was viewed approvingly by clerics and theologians, who were given free rein by the government to attack the international community, regional powers and the United Nations for the devastation of Iraq.[16] Thus, Iraq was undergoing a profound religious revival as the United States and its allies prepared their invasion plans.

Following the invasion, the nationalist rejection of the foreign presence became fused with an Islamist narrative that also excoriated the invasion and occupation. The political vacuum and psychological dislocation occasioned by the invasion and its chaotic aftermath strengthened religious sentiment within the Sunni community. As one Sunni in Baghdad put it: 'During the US invasion, I saw so much chaos and death that I turned to God. Now there is so much corruption and violence that we need an Islamic government according to Shari'a. That would stop a lot of the suffering we have now.'[17] The rise of Islamist feeling within the Sunni community strengthened the power and voice of Sunni clerics and preachers, who began to take an active political role, articulating Sunni grievances, setting out Sunni goals and presenting an Islamist narrative legitimising resistance to foreign occupation.

The political activism of the Sunni clerical establishment had begun in the 1990s under the former regime, as clerics had railed against the sanctions regime and what they saw as the moral and ethical collapse of Iraqi society. But those clerics who dared to attack the regime, whether implicitly or explicitly, were imprisoned or exiled.[18] With Saddam's ouster, however, Sunni clerics began to take charge of the sociopolitical space that had opened up in the political scene. One of the earliest clerics to take an active political role was Sheikh Mahdi Ahmed al-Sumaidi, who was detained in Abu Ghraib prison early in the occupation after a weapons cache was found in his mosque in Baghdad. His distaste for the US presence turned into undisguised hostility after his release: 'Neither the occupation forces nor the government they installed is acceptable', he said. 'The legitimate power is the resistance.'[19] The US occupation, he argued, showed conclusively the necessity of turning to Islam: 'God uses many tools. America's brutality has caused many to understand that Islam is the answer to our problems. The only solution is Islamic government.'[20] Sunni clerics who adopted a Salafist interpretation of Islam began to promote their views among their community, particularly during Friday sermons.

Among them was Sheikh Qutaiba, since 2001 sheikh of the Mosque of the Day of Appeal in Baghdad, who on Saddam's orders issued a call for holy war against the Americans.[21]

Iraq's Salafists believed that Saddam's capture in late 2003 was their opportunity to pick up the baton of resistance from the dominant Ba'athist and nationalist strands. The invasion and occupation had led to a steady increase in recruits. Even members of the former regime were not immune to the ideological attraction of Salafism. In an interview with a French journalist in January 2004, an Iraqi Salafist and former general provided a glimpse into the Salafist worldview. Explaining why resistance to the Coalition invasion had initially been weak, the former general said that 'the Iraqis had had enough of war; that is why the regime collapsed without a fight. This was not their war.'[22] But US actions encouraged resistance. The general then sketched out the main elements of the insurgency. Firstly, there were those (the Ba'athists) who had 'lost all their privileges and would like to restore the former regime or get the Americans to accept their return to political life'. Secondly, there were the patriots who were 'fighting to free their country, either for nationalist reasons or religious motives'. Third were the tribes – 'the Bedouins' – who 'simply want revenge because US soldiers arrested them in front of their wives, confiscated their homes, or killed someone close to them'.[23]

In 2004, according to this officer, it was a fourth strand, the local Salafists, who were on the rise among the ranks of the resistance. The Salafist strand had been in existence before the fall of the regime, but its adherents had been forced to maintain a clandestine presence due to the regime's suspicions of any form of political Islam not under its control: 'In 1990, many soldiers returned to religion. But we were forced to pray in secret. And we could not grow a beard. If we attended the mosques too regularly, we were thrown into prison.'[24] The Salafists were now part of the insurgency:

> Fighting under the flag of Islam ... Muslims in several Arab countries have called for a jihad. We can therefore count on internal and external support. Sheikh Osama Bin Laden has launched an appeal to liberate Iraq. He is a man of principle. If he wants to help us fight our enemy, we will be pleased to accept.[25]

There is a general perception among Iraqi Salafists that al-Zarqawi damaged Salafism when he claimed to be fighting in the name of a purist interpretation of religion. His murderous spree in Iraq and his expansive goals of fighting a global jihad against a wide range of enemies, including Sunni Muslims who did not agree with him (notable in his sponsorship of a suicide operation in Amman in Jordan in November 2005), tarnished the image of Salafism and was seen as a distraction. Over time, Iraqi Salafists began to lose patience with al-Zarqawi's group. This can be seen clearly in the comments of one former military officer, 'Abu Aisha', who adopted the Salafist line. A self-professed Salafist, his rhetoric included a mix of traditional Sunni Arab nationalist suspicion of the Shi'ites and a Salafist excoriation of them as apostates. But Abu Aisha also clearly lacked interest in the international agenda of AQM and its affiliates: 'There is a new jihad now. The jihad is against the Shia, not the Americans. We have been deceived by the jihadi Arabs. They had an international agenda and we implemented it. But now all the leadership of the jihad in Iraq are Iraqis.'[26]

In the aftermath of the invasion, Salafist clerics, teachers and former officers and officials of the defunct regime contributed to the emergence of various powerful Salafist insurgent groups. One of these, the IAI, has said that it is a 'bona fide Salafi group', and believes in a literal interpretation of the Koran and the tradition of the Prophet Mohammed. It has rejected democracy as a 'non-Islamic' system of government, on the grounds that it blasphemously promotes the sovereignty of mankind over God. It rejects any elections that bring to power legislators who will implement 'un-Islamic' laws. The commander-in-chief, or emir, of the IAI has said that the conflict with the Coalition is a religious battle, that the mujahadeen are defending their religion and that their enemies are on the side of evil.

The Iranian presence in Iraq, which has increased dramatically in the past five years, is regarded as a major threat by all the Sunni insurgent groups, including the Salafists. Iran's help has enabled Iraq's Shi'ites to consolidate their hold on the country. The IAI has focused much of its energy on highlighting the Iranian presence. Its commander has criticised Iran for meddling in Iraqi affairs, in particular for its support for Shia militias. The group justifies violence against Shi'ites less on the grounds of theological differences and more on claims that it is defending Sunnis from Shia militias and what it sees as the historical treachery of the Shi'ites. Despite their opposition to nationalism as an ideology, Iraqi Salafists have adopted a mixture of chauvinistic Sunni Arab nationalism and Sunni extremist aversion towards Shi'ites. On 28 December 2006, the IAI commander delivered a powerful anti-Iranian diatribe entitled 'Concerning the Safavid–Iranian Scheme'. The Safavids, an Iranian dynasty, made Shia Islam the official religion of Iran in the sixteenth century, a move seen by the Sunni world, led by the Ottoman Empire, as an affront to and betrayal of the Islamic community. Such sentiments have been revived in the current struggle between Sunnis and Shi'ites in Iraq.

Among other Salafist groups, the 1920 Revolution Brigades describes itself as a 'nationalist jihadist movement' that aims to rid Iraq of occupation and build a country based on the 'lofty principles of Islam which is founded on justice and non-discrimination based on race, ethnic group, or religion'.[27] Another group, the Army of the Mujahadeen (AM), has adopted some elements of traditional Third World anti-colonial rhetoric, and has produced a videotape urging Americans to overthrow their government, saying: 'You have elected these criminals, and thus, you are responsible for their actions … A mafia of the weapons plants represented by Bush, oil companies represented by Dick Cheney and the Zionists by Paul Wolfowitz and Richard Perle hijack the United States of America in an ingenious plan to control the world.'[28] The video concludes by highlighting the negative domestic impact of the George W. Bush administration's policies, arguing that these policies will result in the establishment of a police state in the United States similar to the police states that American governments have traditionally backed in the Middle East.[29]

AQM and its affiliates

In March 2005, AQM's council published the creed and methodology of the movement, in which it expressed its determination to promote and defend *tawhid* and eliminate polytheism. The statement stipulates that anyone who does not believe in the essential unity of God is an infidel and subject to *takfir* (excommunication). AQM believes in the promotion and export of Salafist Islam and subscribes to the view that the Prophet Mohammed is God's messenger for the entire human race. It has a thoroughly hostile attitude towards all kinds of secularism – nationalism, communism and Ba'athism – which it views as a 'blatant violation of Islam', and holds that any individual who supports secularism is to be considered a non-Muslim. Jihad becomes the duty of all Muslims if the 'infidels' attack Muslims and their territories. Like many other Salafist groups, AQM believes that for 'true' Muslims, waging jihad against the enemies of Islam is next in importance to the profession of faith. Finally, the group argues that Muslims – excluding the Shi'ites – constitute one nation. There is no differentiation between Arabs and non-Arabs in Islam; piety is what counts.[30]

AQM views Shi'ites as *rafidah* ('rejectionists') and apostates, and has stipulated that fighting them is more important than fighting non-Muslims. In February 2004, al-Zarqawi called the Shi'ites 'the insurmountable obstacle, the lurking snake, the crafty and malicious scorpion, the spying enemy, and the penetrating venom'. His assault on the Shi'ites was the key factor launching the civil strife between the two sects. Although the

Shi'ites often resisted the temptation to strike back, the bombing of the al-Askariyya shrine in Samarra in February 2006, which destroyed the tombs of two of the 12 imams – descendants of the Prophet Mohammed – was immensely provocative. The attack, widely suspected to have been carried out by al-Zarqawi's group, created a major rift between the Sunni and Shia communities. Following the bombing, outraged Shia mobs, led and incited by Shia militiamen, attacked and took over dozens of Sunni mosques, either in revenge or because they claimed that the mosques had been seized from them by the former regime. Police units stood by as Shi'ites went on the rampage. Throughout most of 2006, Sunni and Shia militias and death squads engaged in what amounted to the ethnic cleansing of mixed areas in a war that was essentially won by the Shia militias, Jaish al-Mahdi in particular.[31] The need to defend their communities from AQM's excesses and to find an ally in this conflict contributed to the gradual but steady reorientation of the Sunni armed groups away from fighting the Americans towards fighting off the Shia militias and death squads and driving AQM out of their areas.

The al-Qaeda leadership outside Iraq may have been sceptical about the tactic of indiscriminate assaults on Shi'ites. In mid 2006, Osama bin Laden issued an audio recording in which he tried to modulate al-Zarqawi's venomous rhetoric. But even after al-Zarqawi's death in June 2006, it seemed that AQM was incapable of diluting its inveterate anti-Shiism. In July 2007, Umar al-Baghdadi, the purported leader of AQM creation the ISI, issued an extensive statement on the Kurds, the Shi'ites and Iran. Although he praised the Kurds for playing an 'honourable' role in Islamic history, he also upbraided them for enabling the Shi'ites to extend their control over Iraq. Kurdish politicians Jalal Talabani and Masoud Barzani had 'strengthened atheist communism and its bitter daughter, secularism'.[32] Al-Baghdadi appealed to the Kurdish people to 'become part of the boundaries of the State of Islam in Mesopotamia'.[33] He had no kind words for the Shi'ites, whom he referred to as the '*rafidah* of Iraq', accusing them of siding with the enemies of Islam throughout their history. The door of repentance was still open to them, however, and they could still return to 'true' Islam. As far as Iran was concerned, al-Baghdadi declared that 'we give the Persians in general, and the rulers of Iran in particular, a two-month period to withdraw all forms of support for the *rafidah* of Iraq and to stop direct and indirect intervention in the affairs of the state of Islam'.[34] Otherwise, they should expect a 'ferocious war', in which they would be destroyed.[35]

CHAPTER THREE

Objectives and Strategy

Each element of the insurgency has its own set of objectives, and strategies for reaching those objectives. There is, however, one objective that all the insurgents – along with the groups involved in the political process – share, namely, the expulsion of all occupying forces. Some insurgent groups have committed themselves to a dual political–military strategy of using force while also offering to negotiate in order to attain the removal of the foreign presence and what they see as the illegitimate political process in place since 2003. AQM and its affiliates do not believe in negotiation, but believe solely in the use of force to expel or destroy the 'infidels' and their supporters. The insurgents also differ among themselves over what should follow the 'liberation' of Iraq.

In October 2006, a jihadist website published an interview with Ibrahim al-Shammari, spokesman for the IAI. In it, al-Shammari defended negotiations, saying that they were 'legitimately permissible … In principle, we are not against negotiating with enemies if the other party is serious'.[1] When insurgent elements have stated that they are prepared to negotiate, they have invariably been referring to negotiations with Coalition forces only, and not the Iraqi government, which they consider nothing more than a puppet of the Americans. The IAI does not accord the government of Prime Minister Nuri al-Maliki any more legitimacy than it did the predecessor government of Ibrahim al-Jaafari, who is also a Shi'ite. Organisations such as the IAI have often sought to open negotiations with the Coalition via the mediation of major Sunni political parties and movements such as the Iraqi

Accord Front, the Iraqi Islamic Party and the National Dialogue, which took part in the December 2005 elections and which have since warily participated in the political process. Some Sunni politicians openly support the insurgents and share their ideological outlook. For example, parliamentarian Husayn al-Falluji, a leading member of the Iraqi Accord Front, has expressed his stance openly: 'We want an Islamic state in Iraq. We have no problem with saying that the resistance brandishes the Islamic flag ... I am in favour of rapprochement among all the forces ... The resistance is first and foremost Islamic'.[2] Al-Falluji reflects the position of mainstream political Sunni groups that argue that the focus of the insurgency should be the US forces in Iraq, and that anyone who helps the Americans is a collaborator and thus a legitimate target: 'The Americans are the number one prey. But those who help the enemy become enemies. War is no picnic. There are two resistances. One is national and the other is transnational. It is the former that is in the majority.'[3]

The Ba'ath Party and some of its affiliates have taken a moderate and conciliatory approach towards negotiation with the occupation forces, though it remains to be seen whether this represents genuine new-found moderation or is merely a tactic. Izzat Ibrahim al-Duri has laid out the following set of 'principles', on the basis of which the Ba'ath Party would be prepared to negotiate:

- Recognition of the resistance in all its forms, to include any group, whether Islamist or nationalist, whose aim it is to liberate Iraq from the invading forces;
- A US announcement of the withdrawal of its forces without restriction or condition;
- The complete cessation by the Coalition and the Iraqi government of raids, round-ups and operations involving killing and destruction;
- The release of all captives, detainees and prisoners;
- The restoration of the pre-invasion army and national security forces.

In an interview in 2006, al-Duri said that he looked forward to a time when Iraqis would be able to build normal and mutually beneficial relations with the United States on the basis of shared interests, arguing that:

> Iraq, like all countries of the world, cannot do without legitimate mutual relations and joint cooperation with America in

all fields of life because of the latter's vast resources, especially in the economic, technological, and developmental spheres. We understand the role and strategic interests of America as a great power.[4]

He added that such relations must be forged on the basis of freedom, independence and the right of the individual to choose his own way of life.

The objectives of the nationalist-Islamists show some similarities with those of the Ba'athists. However, the former are not interested in restoring to power the previous regime or its surviving leadership, whom they blame for many of the ills that have befallen Iraq over the past three decades.

The most extensive articulation of Iraqi Salafists' objectives and strategies has been undertaken by the larger groups, including Ansar al-Sunna, the IAI and the 1920 Revolutionary Brigades. The main objective of Ansar al-Sunna is to expel the Coalition and turn Iraq into a Sunni Islamic state. According to the group's founding commander, Abu Abdullah Hasan Bin Mahmud, the struggle in Iraq is the beginning of a larger struggle to establish the 'rule of Allah on earth', and marks the beginning of the end for 'tyranny and apostasy'.[5] While most Iraqi Salafist movements have not gone beyond the main objective of establishing an Islamic state in Iraq, there is clearly a transnational aspect to Ansar al-Sunna's thinking, suggesting an ideological affinity with transnational Salafists.

The IAI's strategy has been clear and consistent since 2005. The only permissible political system is one that adheres to Islamic laws. The IAI opposes policies that might be viewed as extremist or fanatical. Thus, individuals may not be subjected to *takfir* without the agreement of all Salafi scholars, and in *takfir* cases, the group distinguishes between those who actively support enemy forces and those whose circumstances mean that they have to deal with them on a daily basis. IAI mujahadeen must avoid acts that damage their reputation and that of the jihad. The jihad must not succumb to hubris, and power-hungry individuals must not be allowed to join the leadership. The group's leader must have a *shura* (consultation) council to advise him on policy, religious and operational matters. Emphasis is placed on the importance to a resistance movement of flexibility and awareness of the group's own strengths and weaknesses. Financial independence and stability are also seen as critical foundations. The IAI cites the maintenance of secrecy and strict discipline among the rank and file as further characteristics of a successful resistance group. The organisation permits suicide operations, provided that they are approved by the commander, their benefits are clear, they can be expected to weaken the enemy and their commission will not cause greater harm to Muslims. The group's political programme calls for the establishment of God's rule in Iraq, but its goal is not to establish a doctrinaire government that interferes with other Islamic sects. The IAI is dedicated to establishing Iraq as a major Islamist power with a welfare system based on Islamic principles of the fair and equitable distribution of wealth. What is particularly noteworthy in the IAI's enunciation of its political principles is that it does not express an interest in establishing a caliphate, or Islamic system of government incorporating all Muslim territories.[6] The IAI has conducted a two-pronged political–military strategy, combining offers of negotiation with military operations. It has been in talks with US officials since May 2006, reportedly mediated through Iraqi President Jalal Talabani's office and Vice-President Tariq al-Hashimi's Sunni Tawafuq Front.[7]

The 1920 Revolutionary Brigades organisation has set out its objectives and strategies in some detail. As part of the General Command of the Islamic Resistance Movement, a group of like-minded insurgent organisations, it has pledged to continue 'blessed jihad' to end the occupation in all its forms. In January 2007, it stated that:

- its goal at this stage was 'to fight the invader who occupied our land';
- the jihad in this phase was defensive, not offensive (this was in context of a broader debate within political Islamic circles over 'defensive' and 'offensive' jihad);
- its operational method was an 'all-encompassing jihad', that is, incorporating military, political and indoctrination dimensions;
- it prohibited the killing of Muslims and non-combatants regardless of religion, sect or nationality – 'the basis of our jihad is not to cause harm to civilians' – and did not believe that ends justified means;
- it demanded discipline and obedience from its members;
- the basis of its jihad was 'to investigate the evidence, and to interpret it on the basis of Islamic jurisprudence and its accepted principles', and that it supported *ijtihad*, the process of making legal decisions through independent interpretation of legal sources, the Koran and the Sunna (traditions established by the Prophet).[8]

In June 2007, one of the group's senior officials, Mujahid Abd-al Rahman, stated in a Internet discussion that the main objectives of the 1920 Revolutionary Brigades were 'the liberation of our country from the infidel occupier, pursuing and crushing all his henchmen and collaborators', and the establishment of an Islamic state based on the Koran and the Sunna. Abd-al Rahman claimed that, over the four years of occupation, the insurgency had defeated the 'entire Zionist-American enterprise', and that 'colonialist' US ambitions to divide and destroy the Arab world beyond Iraq had come undone on the battlefields of Iraq due to the resistance of the mujahadeen.[9]

AQM has provided extensive documentation of its objectives, notably in its magazine *Dhurwat al-Sanam*. The magazine's first issue carried an article entitled 'Who Are We?'. In it, AQM defined its mission as fighting the 'infidel invaders' until they were expelled from Iraq. Their expulsion was to be followed by the establishment of an Islamic state, and then by a caliphate throughout the Islamic world. According to Abu Maysara al-Iraqi, one of AQM's leading spokesmen, the organisation's goals are:

- to remove the 'aggressor' from Iraq;
- to affirm *tawhid*, or oneness of God, among Muslims;
- to propagate the message that 'there is no god but God' in those countries where Islam does not reach;
- to wage jihad in the cause of God in order to exalt his word, liberate all Muslim territories from infidels and apostates, and establish sharia law in these territories;
- to support Muslims everywhere, restore their dignity – which the invaders and traitors have desecrated – reassert their usurped rights and improve their general situation; and
- to establish a wise caliphate similar to the theocracy established by the Prophet Mohammed.[10]

Between late 2006 and mid 2007, the ISI developed objectives of its own, which were essentially an elaboration of those of AQM and its umbrella organisation, the Mujahadeen Shura Council. According to Uthman Abd al-Rahman, the head of the ISI's Sharia Commission, the ISI's primary objectives are to 'spread monotheism on earth, cleanse it of polytheism, to govern according to the law of God, to repulse the aggressors'.[11] It is thus clear that the goals of AQM, the Mujahadeen Shura Council and the ISI extend well beyond Iraq.

CHAPTER FOUR

Organisation, Targeting, Operational Art and Tactics

Open-source information about the internal structures and organisation of Iraq's insurgent groups is often unreliable. The most successful insurgent groups have implemented effective counter-intelligence procedures to prevent penetration. Members of the insurgency are, unsurprisingly, reluctant to provide information about the structure and organisation of their groups and how they were recruited. Moreover, even in the case of a relatively large group such as the IAI, individual insurgents know little about other cells within their own organisation. However, considerably more information is available on targeting, operational art and tactics.

Internal structures and organisation

Looking at the structure and organisation of Iraqi insurgent groups, two major characteristics stand out. The first is that the vast majority of Iraqi insurgent organisations are hybrids, in that they consist of a mix of hierarchical and decentralised structures.[1] No Iraqi insurgent organisation or group is wholly hierarchical, not even the previously rigidly hierarchical Ba'ath Party. Such a structure would make a group susceptible to penetration by hostile forces and to leadership decapitation. At the same time, none is completely decentralised and without a leadership structure. Such an organisation would not be able to effectively undertake operations or exercise command, control and coordination. Thus the IAI, for example, has branches and committees – such as its Sharia Committee and Higher Judicial Committee – and military and political bureaus. Its military units appear to be sophisticated, and include the following specialised types: engineering support for building and placing improvised explosive devices (IEDs), combat groups with light and medium weapons, mortar and missile units and surveillance units, 'special services' (tasked with eliminating Badr Brigade and Jaish al-Mahdi personnel) and a sniper unit. The IAI also claims to have a 'legitimacy committee', whose task it is to determine the legality and morality of proposed military actions. Targeting is supposedly selective and rigorous: the IAI claims that it does not target oil and other critical infrastructure, because of the harmful effects such attacks would have on the general population. The IAI conducts 'martyrdom' operations, but under certain rules: such operations are conducted when they are the only possible way to attack the enemy; the attacker must not be a scholar, a military commander or other essential member of the organisation; and attacks should be motivated by the desire to defend Islam, and not by despair.[2]

The second structural characteristic of Iraqi insurgent organisations is functional specialisation. The largest groups often possess a large cadre of personnel with specialised skills, including bureaucratic skills, and adequate funding, which gives them both the ability to undertake a wide range of simple and complex operations and nationwide reach. Such groups are able to create a wide range of specialised combat and combat-support cells. The Ba'athists and their affiliates are able to show a high degree of functional specialisation because they previously controlled the state and its institutions, some of which they recreated in a decentralised form in order to conduct their insurgency. The Salafist Al-Fatihin Army also seems to have developed an extensive and highly specialised organisational structure, perhaps based on that of its parent organisation, the IAI. Al-Fatihin claims to have a wide variety of internal institutions and councils ('Leadership', 'General Command', 'Shura', 'Sharia', 'Military' and 'Media'). Its Military Council is alleged to control ten combat brigades operating in cities in central and northern Iraq.[3]

AQM also developed an advanced level of functional specialisation, which it passed on to the Mujahadeen Shura Council and the ISI. In addition to combat battalions, the group's military wing has a security and reconnaissance unit which vets new members, collects intelligence and recruits agents in the Iraqi security forces, contractor companies and the Coalition's transport and logistics-support organisation. The organisation's suicide battalion, Al-Barra Bin Malik, led by Abu Dajana al-Ansari, was originally mostly made up of foreign Arab volunteers, but by 2006, an influx of Iraqi volunteers had begun. AQM's Sharia Council, headed by Umar al-Baghdadi, was responsible for conducting research and for organising the dissemination of the group's ideology and beliefs, including through the publication of its magazine, *Dhurwat al-Sanam*. The group's 'information ministry' under Abu Maysara al-Iraqi released statements, bulletins and video and audio tapes, and became skilled at placing its products on the Internet. The 'finance ministry' received donations from sympathetic businesses and from mosques, particularly in collections following Friday prayers. (Funds for AQM have also been sourced from other Arab countries, including Jordan, where the authorities have dismantled several groups collecting money for al-Zarqawi.)

Those groups that cannot afford a wide range of specialised cadres may be forced to 'borrow' them from other groups. Such groups – the tribal insurgents and small localised brigades – are unable to undertake a wide range of operations, and may be focused on defending their local area, whether from Coalition forces, the Iraqi security forces, Shia militias or, increasingly, AQM and its affiliates, rather than on fighting in a nation-wide insurgency. For many, this is not a particularly onerous constraint, as they do not aspire to do more.

Targeting, operational art and tactics

Iraq's insurgents have a wide range of potential targets. Firstly, there is the Coalition and its associated networks and infrastructure, including supply convoys and private security companies. Attacks on convoys protected by private companies increased significantly in line with the growing US dependence on armed civilian guards – there were 869 such attacks between June 2006 and May 2007, three times as many as during the preceding 12 months.[4] This category also includes the foreign companies and workers who flooded into Iraq in the wake of the invasion to participate in the expected windfall from reconstruction, development and the provision of services to the enormous Coalition military and civilian presence. The second set of targets comprises those Iraqis who work with or for the Coalition and the Iraqi government. This is a vast category, including public officials and bureaucrats, teachers, doctors, intellectuals, professionals, translators, police and military personnel. The third set of targets consists of ordinary Iraqi civilians. Some insurgent groups have chosen to attack all these sets of targets, while others focus on one or more. Some groups have chosen not to attack a particular set or sets of targets, and have criticised other insurgents for, for instance, attacking the Shi'ites as a community, or targeting Iraq's oil and other critical infrastructure.

Iraq's Sunni insurgents have used three types of tactic and operational style: terrorism, guerrilla warfare and anti-infrastructure operations. These are discussed in detail below.

Terrorism

There is a distinction between insurgent groups that use terrorist tactics and terrorist organisations as such. Insurgents tend to resist being called terrorists, and are generally reluctant to use terrorism on an extensive scale. Many use it during the opening stages of an insurgency, when they want to eliminate rivals, to show the government and populace that they are serious and to provoke reprisals from the government. Usually, however, the use of terrorism declines as an insurgency gathers momentum. The Sunni Iraqi insurgency has proved different in this regard. Rather than decreasing, terrorism seems to have increased as this insurgency has gone on. There are groups within the insurgency whose entire modus operandi has been based on the use of terror through suicide operations, assassinations of individuals or groups associated with or collaborating with the Coalition, and the kidnapping and execution of foreigners. Suicide operations have become a major speciality of some insurgent groups. No other conflict arena in which suicide bombings have been used (Lebanon, Palestine or Sri Lanka) has witnessed such frequent attacks or as many deaths from such attacks as Iraq. Suicide attacks have been used on a massive scale against civilian targets, often Shia civilians, but also by Sunni religious extremists against other Sunnis. Such attacks have also often targeted the Iraqi security services. Most of Iraq's suicide bombers are foreigners, usually from Saudi Arabia and other Gulf states, and insurgent groups have created sophisticated Internet networks to help recruit suicide bombers from abroad.

Suicide car bombs, known in the US military as 'vehicle-borne IEDs', have emerged as one of the insurgents' most effective weapons, along with roadside IEDs. They are often directed against targets such as police stations, recruiting centres for the security services and US convoys. They deliver a large amount of firepower and inflict large numbers of casualties at little cost to the attacker organisation, lowering the morale of Coalition and government troops.

Attacks on civilian targets began in earnest in August 2003, and have steadily increased since then. They have included the assassination of Iraqis cooperating with the Coalition Provisional Authority and the Governing Council, and suicide bombings targeting the United Nations headquarters, the Jordanian Embassy, Shia mosques and civilians, the Red

Cross, Kurdish political parties, the president of the Governing Council, hotels, churches, diplomats and restaurants. Iraqi police and security forces are targeted in ambushes and execution-style killings. Insurgents have also attacked private contractors working for the Coalition, and assassinations of local and government officials, translators for Coalition forces, employees at Coalition bases, informants and other 'collaborators' have been common. Assassinations have taken a variety of forms, from close-range small-arms fire and drive-by shootings to suicide car-bombers ramming convoys. Kidnapping emerged as another insurgent tactic from April 2004. Foreign civilians have borne the brunt of this particular trend, though US military personnel have also been targeted. After kidnapping a victim, insurgents typically make some sort of demand of the hostage's government, and set a time limit for the demand to be met, often 72 hours. Beheading is often threatened, and several victims have been killed in this way. In some cases, videotapes of beheadings have been distributed for propaganda purposes. However, 80% of hostages have been released. The goal of kidnapping appears mainly to be to terrorise foreign civilians, attract media attention, and possibly to inspire recruits. Almost all kidnappings have been conducted by radical Sunni groups on the fringes of the insurgency.

Guerrilla tactics
Iraqi insurgents use the traditional guerrilla tactics of hit-and-run attacks, raids, ambushes, car bombs, IEDs and assassinations. The insurgents usually operate in small teams of five to ten men, which facilitates mobility and reduces complexity in command and control and the likelihood of detection. However, operating in small teams also reduces the amount of firepower that can be brought to bear against Coalition forces. Initially, insurgents would engage bands of militants in set-piece firefights with better-armed and trained US combat forces, in which they usually lost badly. Gradually, in a process of 'combat Darwinism', the survivors of such engagements learned, adapted and became more technically flexible. Insurgent groups improved their training and issued tactical manuals, which are often distributed on the Internet where other groups can access them.

Over time, the insurgents developed the capacity to launch larger, more complex and better-executed attacks involving as many as 150 fighters, as for example in and around the city of Ramadi in Anbar province, where complex attacks were led by former Iraqi Army commanders. In April 2005, two particularly audacious attacks by large insurgent forces were launched, against Abu Ghraib prison south of Baghdad and against a US Marine base at Husayba. In the first attack, two separate columns of insurgents assaulted Abu Ghraib following an accurate and sustained bombardment of US positions by 80mm and 120mm mortars and two vehicle-borne IED assaults aimed at breaching the prison walls. In the second case, a group of 100 insurgents launched a well-coordinated and complex attack on the Marine outpost.[5] Guerrillas in Diyala province have also become increasingly well organised and trained, and in 2006, the province was the third most dangerous place for US troops in Iraq after Baghdad and Anbar. This was illustrated by a particularly ferocious assault in the town of Muqdadiya in March 2006, when a 100-strong insurgent force attacked a court building to release detained insurgent suspects, holding off US and Iraqi reinforcements. Seventeen security personnel were killed, several police cars destroyed and the court building and adjacent police station set alight; 33 detainees were released.[6] In another example of insurgent capabilities, an insurgent group placed 27 IEDs along a one-mile stretch of road in the town of Buhriz. For each real device they also laid three or four decoys, which slowed down patrols, giving the insurgents time to launch coordinated attacks with rocket-propelled grenades, mortars and machine guns. The skills of insurgents in Diyala may also derive from the fact that Baqubah, the province's main city, is home to many Ba'ath Party loyalists and military and intelligence officers from the former regime.[7]

The ability to learn and adapt is also evident in the efforts insurgents have made to counter US attempts to defeat the IED threat. First-generation IEDs were relatively small and simple, often a 155mm or 152mm artillery shell hidden in a wall or embankment beside a road. Insurgents would run wires from the device to a hand-held trigger, which they could activate from a nearby hiding place. As US troops devised counter-measures to detect these IEDs, by spotting the wires or looking for suspicious individuals nearby, insurgents changed their approach and began using remote triggers such as garage-door openers and mobile phones to detonate the devices from greater distances. They also began using more powerful explosives, sometimes connecting multiple artillery rounds to boost destructive force, and by March 2004 insurgents were 'daisy-chaining', emplacing multiple 155mm rounds in a row. In early 2004, US troops started receiving their first large-scale deliveries of jamming devices designed to block the wireless signals the insurgents were using to set off IEDs. The insurgents adapted faster, however, burying IEDs under roads so that they would blast up through the thin floors of armoured Humvees, and switching to hard-wired devices or pressure-plate IEDs, which explode when a vehicle goes over them. By 2007, they were adapting again and using 'explosively formed projectiles'.[8]

Iraq's insurgents have shown great inventiveness, flexibility and adaptability in their tactics and operations. Particularly remarkable too is the unprecedentedly high operational tempo they have achieved in the areas of car bombings, suicide operations and IED use.

Though most of the techniques they have used have not been original, Iraq's insurgents have been responsible for one innovation: the attempt to gain a chemical-warfare capability. The first indication of interest in this area came in March 2004, when the Iraq Survey Group (ISG)[9] began investigating a group of insurgents known as the al-Aboud network, which had begun seeking a chemical-weapons capability in late 2003. Although the ISG believed that the al-Aboud network was not the only group planning or attempting to produce or acquire chemical agents, it focused on the group because of the maturity of its chemical-weapons production, and the severity of the threat posed by its weaponisation efforts. The ISG created a team of experts to investigate and dismantle the network. By June 2004, it had identified and neutralised the chemical suppliers and chemists, including former regime members, who supported the network. A series of raids, interrogations and detentions disrupted key activities at al-Aboud-related laboratories and safe houses, but the network's leaders eluded capture.

Al-Aboud had already established links with insurgent groups interested in acquiring a chemical-warfare capability. In the stronghold of Falluja, insurgents grouped around Jaish Muhammad recruited an inexperienced Baghdad chemist to develop chemical agents, including tabun and mustard gas. Jaish Muhammad and other Falluja-based insurgents planned to use chemical weapons against Coalition forces. If the insurgents had acquired the necessary materials, fine-tuned their production techniques and gained a better understanding of the principles behind effective dispersal, the consequences for the Coalition could have been devastating.

In general, the availability of chemicals and materials and the expertise derived from the previous regime's weapons-of-mass-destruction (WMD) programmes means that the threat to Coalition forces, Iraqi security forces and Iraqi civilians from possible chemical and biological agents has remained high.[10] In October 2006, AQM leader Abu Hamza al-Muhajir issued an audio statement in which he emphasised the need for people with scientific and technical skills to join the mujahadeen in Iraq, telling potential recruits that 'the battlefield will accommodate your scientific aspirations'. AQM detonated trucks carrying explosives and chlorine gas three times between late January and late February 2007. By the end of April 2007, eight chlorine-gas attacks had taken place.[11] Though several people were killed in the attacks and many more were injured, the impact of the attacks was mainly psychological, raising fears that worse was to come. There have not, however, been any further chemical-warfare-related attacks thus far.

Attacks on infrastructure

Some groups in Iraq have specifically targeted the infrastructure that both the Coalition and the Iraqi government need in order to implement reconstruction and development programmes. In their attacks on bridges, schools, police stations, post offices, roads, electricity pylons and oil pipe-lines, the insurgents' goal is to prevent the government and the occupying forces from bringing stability, security and law and order to Iraq. If the Coalition and the Iraqi government fail in their reconstruction and development endeavours, they will lose legitimacy.

Former elements of the Ba'ath Party were the first to target infrastructure, with a particular emphasis on the oil infrastructure:

> Iraqi oil ... will be a legitimate and a permanent target of the armed resistance plans to liberate Iraq and defeat the invaders ... the armed resistance will use every possible means militarily and technically to prevent the occupier from stealing Iraq's oil and us[ing] its revenues with anyone, under any circumstances, on the national and international levels ... On this basis, everyone who collaborates with the occupier, such as employees, merchants, middlemen, whether Iraqis, Arabs or non-Arabs, will be watched and targeted without any hesitation.[12]

Other groups that have targeted Iraqi oil include AQM and its creation, the ISI.[13] In December 2004, Osama bin Laden issued the following statement:

> One of the most important reasons that made our enemies control our land is the pilfering of our oil. Exert all that you can to stop the largest stealing operation that [has taken] place in history ... Be active and prevent them from reaching the oil, and mount your operations accordingly, particularly in Iraq and the Gulf.[14]

In late 2005, senior al-Qaeda figure Ayman al-Zawahiri exhorted the mujahadeen in Iraq to concentrate their campaign on the 'Muslims' stolen oil, most of whose revenues go to the enemies of Islam'.[15] The anonymous jihadist author of *Jihad in Iraq: Hopes and Dangers* emphasises the importance of Iraqi oil to the occupation and its reconstruction effort. Striking at oil facilities is the best way of reducing oil exports from Iraq and of 'forc[ing] the US, in the event it stays in Iraq, to pay by itself for the cost of living of the Iraqi people on top of the cost of the occupation itself'. Oilfields, pipelines, extraction areas, ports and refineries are all considered critical targets.[16]

Not all Sunni Iraqi insurgent groups have shown such an interest in attacking infrastructure. Many nationalist-Islamist groups have argued that such a tactic is counterproductive, as it alienates ordinary Iraqis by depriving them of resources and services that the insurgents themselves are often not capable of providing effectively. Moreover, such attacks make reconstruction following the 'defeat' and withdrawal of the occupation a more onerous task.

The success of the Sunni insurgent groups in using a variety of terrorist and guerrilla methods to create disorder and sustain a high level of violence up to the end of 2006 was not matched by a parallel effort at the political level to define and articulate visions or goals. This failure stemmed from a number of significant structural weaknesses within the insurgency itself. These weaknesses are the subject to which we now turn.

CHAPTER FIVE

The Insurgency's Internal and External Problems

The insurgency in Iraq is largely Sunni Arab, despite the fact that Sunni Arabs only account for around 17% of the population. Partly as a result of this, the insurgency suffers from severe weaknesses. It has never been able to transcend the parochial and particularistic interests of Iraq's various communities. Even when the country appeared to be on the verge of a genuine national uprising in 2004, when elements of the Shia population mobilised under the leadership of cleric Moqtada al-Sadr, the potential for unity was less than many hoped, despite the claims of some Iraqis that, at last, their country was reacting in a unified national way to the foreign presence.[1] As became clear, al-Sadr was not rising in support of the Sunnis, nor was there any coordination with Sunni insurgent groups except at the very lowest tactical levels.

The insurgency's lack of unity, reflected in the existence of myriad groups, has had a negative impact on mobilisation, organisational coherence, the promotion of an effective ideology of resistance and the articulation of goals to which all Iraqis could subscribe. Moreover, the insurgency has not succeeded in attracting effective international support, nor have its members been able to find sanctuary abroad.

Not a national war of liberation

The insurgency in Iraq is not a national-liberation struggle in the traditional sense of a war against an occupying power. It may be perceived as such by much of the Sunni community, but the insurgency has failed to move beyond its ideological parochialism to incorporate other Iraqis. Indeed, it has contributed, along with other elements in the country, to sectarian divisions and thus to national disunity. There is extensive active and passive support for the insurgency within the Sunni Arab community, support which has increased since the emergence of the Shi'ites as the major political force in the country[2] – though it is important to remember that not all Sunnis have taken up arms, and many are involved peaceably in the political process.

None of Iraq's other ethno-sectarian communities is as heavily involved in the insurgency. Sunni Kurds make up around 18% of the country's population and Shia Arabs constitute roughly 60%. The remaining 6% comprises Turkmen, Christians and other groups. These figures are vigorously disputed, particularly by Sunni Arabs and Turkmen, many of whom are convinced that Shi'ite numbers are inflated. Sunnis often complain that the Coalition exaggerates the demographic weight of the Shi'ites for political reasons, and many Sunni Arabs claim that they constitute a majority, sometimes boosting their community's numbers by including Kurds and Sunni Turkmen in their count. However, a supposed demographic superiority is not the Sunnis' main argument for why they ought to have preponderant power. The claim that the Sunnis 'built' Iraq, and that the country therefore 'belongs to' them is a popular one in Sunni circles in Iraq.[3]

Very few Kurds have taken up arms in support of the insurgency. For the great majority of Kurds, the US invasion represented liberation from the ever-present threat posed to their region, autonomous since 1991, by Saddam's regime. For many Kurds, particularly those who grew up without any Iraqi control, as well as those whose lives were seared by atrocities during the Saddam years, the Coalition presence was a step towards the fulfilment of the Kurdish nationalist dream of an independent state, even as the obstacles to that state remained seemingly formidable. For other Kurds, the US invasion was a step towards the creation of a federal Iraqi state in which the Kurds would play a key role in Baghdad, precisely with a view to guaranteeing the eventual emergence of such an entity.[4]

There are, however, two groups of Kurds that are or could be prone to supporting the insurgency. Firstly, there are those Kurds who take a primarily Islamist view, in particular those associated with the Islamist Ansar al-Islam group and Ansar al-Sunna, its largely Sunni Arab spin-off.[5] Secondly, there are those Kurds who have remained tied to the remnants of the former regime through patronage, and by virtue of the fact that they have lived for generations in largely Arab areas such as Baghdad, which is home to up to a million Kurds. Kurdish security forces have expressed concern over the increase in the activities of Islamist insurgent groups in the Kurdish region; many of these militants are members of al-Qaeda's Kurdistan Brigades. The Kurdistan Brigades have also moved into bases abandoned by Iranian opposition movement the Mujahadeen-e-Khalq within Iraq proper, a few miles from the borders of the autonomous Kurdish region. It is also alleged that a small number have moved into the Kurdish towns of Kalar and Darbandikhan, south of Suleimaniya.[6]

The Shia Arab response to the foreign presence in Iraq has been the most complicated. Roughly two-thirds of the Ba'ath Party's low- and middle-ranking personnel were Shi'ites, and the regular Iraqi Army contained a large number of Shia junior and middle-ranking officers. Many Shi'ites linked to the former regime fought in the insurgency against the Coalition in the early days, before it became more Sunni Islamist and anti-Shia. In general, most Shi'ites viewed the overthrow of the former regime as a well-deserved liberation from a tyrant who had oppressed them and kept the southern region, where they constituted the majority, under tight control and deprived of social services. But Iraqi Shi'ites tend to be Iraqi nationalists, and many of the political and social movements that emerged from this community in the days following the collapse of the regime viewed the Coalition with considerable suspicion. They did not believe Coalition rhetoric about 'liberation' and 'democracy'. Many remembered only too well the previous international coalition's failure to save them from Saddam's army in 1991, when they rose in rebellion in the south following Iraq's defeat. Following the 2003 invasion, major Shia political groups chose to oppose the Coalition presence through political rather than military means, a decision in which senior clerics played a major role. Groups such as the Supreme Assembly of the Islamic Revolution in Iraq of Ayatollah Mohsen al-Hakim, the al-Da'wa party and others viewed the Coalition presence as a necessary evil that they could use to empower themselves politically.

Some Shia groups have fought against the Coalition both in the centre of the country, largely against American forces, and in the south, against the British. The key player here has been Moqtada al-Sadr's Jaish al-Mahdi. Al-Sadr received assistance from Sunni insurgent cadres, who trained his militia in tactics and the handling of heavier weapons, including rocket-propelled grenades and mortars. This ad hoc cooperation between Sunni and Shia insurgents boosted the morale of anti-Coalition elements within Iraq for a time, prompting many to compare the Sunni–Shia insurgency of 2004 with the joint Sunni and Shia nationalist uprising against the British in 1920. Despite these contacts, however, Sunni–Shia cooperation did not persist. Al-Sadr and the Sunni insurgents shared a common anti-Coalition goal, but al-Sadr had his own political agenda, which did not include politically and strategically re-empowering the Sunni Arab community. Furthermore, when extreme anti-Shia jihadist elements began targeting Shia civilians and places of worship, al-Sadr felt compelled to retaliate against Sunnis, even as he decried the descent into sectarian strife.

There is no clear data on the role of Christians in the insurgency. Many Christians were supporters of the former regime during its first two decades because it tightly controlled manifestations of Islamist revivalism, and because it provided them with considerable opportunities for advancement. Some Christians, particularly from Baghdad and Mosul, areas where Christians are concentrated, probably did take up arms against the Coalition, but tens of thousands more have left Iraq in response to the rise of Islamist politics – both Sunni and Shia – since the fall of Saddam's regime.

The Turkmen minority has played an important role in post-Saddam Iraq, despite its small size.[7] While the majority of Turkmen in Iraq are Sunni, the community also has a substantial Shia minority. During 2005, there was extensive violence between Sunni Turkmen militias and their Shia counterparts in the poverty-stricken city of Tel Afar in Nineva province in the northwest, violence that, according to local Sunni Turkmen, was supported by the government in Baghdad, Shia militias and Iran. The Sunni Turkmen of Tel Afar had benefited immensely under the former regime, serving faithfully in the Ba'ath Party apparatus, the security services, the public sector and, above all, the army. They reacted violently to the loss of material privileges and the threat to their identity posed by the changes that followed the regime's fall. The Turkmen have also come into conflict with the Kurds, particularly in the town of Kirkuk, which the Kurds claim as their own, and which they want to make the capital of their own region.[8] Issues pertaining to the Turkmen have attracted the attention of neighbouring Turkey, largely due to the efforts of the Iraqi Turkmen Front, whose membership is made up largely of middle-class Turkmen from Kirkuk and Baghdad.[9]

Factionalism and rivalry

As we have seen, the Sunni insurgency is fractured and incoherent in its strategies, ideologies and goals. Relations between the five strands of the insurgency outlined in Chapter 1 have often been characterised by profound differences over ideology, objectives and methods, as well as over policy towards the Coalition, the Shia-dominated government and the mainstream Sunni political parties. There have been three distinct splits within the insurgency. The first occurred within the ranks of the 'mainstream' insurgency, largely pitting Ba'athists and nationalists against Iraqi Salafists. The second pitted mainstream insurgents (including Iraqi Salafists) against the transnational Salafists. The third, concurrent with the split between mainstream insurgents and the transnational Salafists, occurred between the tribes and the transnational Salafists.

Factionalism and rivalry within the mainstream insurgency
The Ba'athist and nationalist elements within the Sunni insurgency have had profound ideological disagreements with the Iraqi Salafists, despite partially sharing a common worldview and a past record of cooperation. The 1920 Revolution Brigades organisation has shown nothing but disdain for the Ba'ath insurgent group led by Izzat Ibrahim al-Duri, and has disavowed any connection with the Ba'ath Party. Meanwhile, in April 2007, there was a major verbal confrontation between the Ba'ath Party and the Islamist AM. The row began when a Ba'ath spokesman claimed that the Ba'ath Party had been the original brains behind the insurgency, and that it remained a key element within it. The spokesman also said that the Ba'ath Party cooperated with the IAI and the AM, but that it had no links with AQM or its affiliates:

> [Al-Qaeda] has a methodology, a vision, and a strategy that differs from the methodology, vision and strategy of the Ba'ath and the rest of the other nationalist resistance factions, so, from our perspective as Ba'athists and resistance factions, there is absolutely no form of any relationship whatsoever with Al-Qaeda.[10]

The claims prompted quick responses from both the IAI and the AM. The former issued a statement denying any 'relationship or coordination at all between the Islamic Army and the so-called Ba'ath'.[11] The AM's leaders stated that the Islamists had opposed the occupation from the very beginning, whereas the Ba'athists had surrendered the country to foreigners and then gone into hiding, leaving no presence in the field. The AM also stated emphatically that there was no ideological compatibility between the Ba'ath and the combatant Islamic factions, and that the latter 'considered the Ba'ath's doctrine an infidel doctrine that removes its adherents from the community of the faith'.[12] The AM's attack prompted a counter-response from Salah al-Mukhtar, a former regime official writing on an insurgent website. Al-Mukhtar praised the presence of the Ba'ath Party in the field of battle, and argued that the Ba'ath insurgency had actually helped Islamist groups to emerge and thrive.

Exasperation with factionalism and rivalry within the mainstream insurgency have prompted many groups to seek closer cooperation and coordination among themselves. In April 2007, some of the smaller insurgent organisations announced the formation of the Iraqi Resistance Popular Front, and called upon others to join it in the 'battle for liberation and to build an independent state'. In the same month, another 'front' was formed in Anbar province, under the umbrella of the IAI. Meanwhile, nine other insurgent groups established a 'National Resistance Coordination Bureau' in an attempt to unify the resistance and put an end to factionalism and infighting.[13] Participating groups included Ansar al-Sunna, the Muslims' Army, the Army of the Naqshbandi Way, the Islamic Front for the Iraqi Resistance, the 1920 Revolution Brigades, the al-Faruq Brigade, the Mustafa Battalion and the Ansar Allah battalion. In May 2007, the IAI, the AM and Ansar al-Sunna formed the Joint Resistance Front (JRF), calling on the 1920 Revolution Brigades to join them (it did not do so). The new group seemed particularly keen to distinguish itself from AQM. Its founding statement announced the goal of repulsing the enemy without the use of 'outrageous or excessive acts', and said that its military operations were focused on the occupiers and their 'puppets', not Iraqi civilians. The JRF also stressed that it would avoid tarnishing the image of the jihad and the mujahadeen, and would not be diverted by internecine battles from the main battle with the Coalition and the Shi'ites. This seems to have been a criticism of the conflicts between AQM and its affiliates and various Iraqi insurgent movements.[14]

Much of the impetus for the formation of groups such as the JRF came from a belated recognition among like-minded insurgents that the insurgency's energy in the military arena was not being matched by similarly determined political efforts. A true insurgency, as many commanders in Iraq began to realise, required political and military wings acting in parallel.

Disputes between the mainstream insurgency and AQM
The often bitter disputes between Ba'athists, nationalist-Islamists and Iraqi Salafists (that is, the 'mainstream') pale in comparison with the tensions between these groups and the transnational Salafists grouped around AQM and the ISI. Strains emerged as early as 2004, in the insurgent bastion of Falluja. As they prepared for the American assaults on the city, it appears that local insurgents increasingly came to resent the prominence and high-

profile role adopted by the transnational Salafists, most of them non-Iraqis following al-Zarqawi and his deputy, Anas al-Shami.

Al-Zarqawi is at the root of the divisions between the mainstream insurgents and AQM. From late 2005, AQM's focus on attacking Iraqi police and army recruits and its efforts to promote splits between Sunnis and Shi'ites became increasingly incompatible with the aims and tactics of the mainstream Sunni insurgent groups. Al-Zarqawi's rigid and inflexible ideological stance, particularly towards the Shi'ites, his use of Islam to justify extreme violence against both the Coalition and the Shi'ites and his general modus operandi began to create tensions, and mainstream Iraqi insurgents attacked him for what they saw as his mistaken understanding of Islam, and for misusing religion to justify AQM's violent acts. It is possible that the Mujahadeen Shura Council, created in January 2006, was set up to curtail al-Zarqawi's activities. Although few groups accepted the Mujahadeen Shura Council's invitation to join the umbrella organisation, the response was not overly hostile. However, when the Mujahadeen Shura Council set up the ISI in October 2006, the move was viewed with consternation by Iraqi insurgent groups not affiliated with AQM, as well as by Sunni politicians in Baghdad. Groups such as the IAI and the Al-Fatihin Army opposed the step, arguing that the time was not right to create an 'Islamic State of Iraq'.

Major Sunni insurgent groups began cutting their links with AQM. In the Sunni heartlands, mainstream groups accused AQM of killing, kidnapping and torturing dozens of their fighters, clerics and followers. The General Command of the Iraqi Armed Forces, a small Ba'athist offshoot, split from AQM in September 2006. Abu Marwan, a spokesman for the General Command, outlined the immediate reason for the split as follows: 'al-Qaeda killed two of our best members, General Mohammad and General Sa'ab, in Ramadi, so we took revenge and now we fight al-Qaeda'.[15] The struggle between the Ba'ath and AQM grew increasingly bitter, with fierce firefights between the Ba'athist al-Awda group and AQM jihadists in western Anbar. Meanwhile, AQM gunmen distributed flyers threatening to execute anyone from the al-Awad tribe, which constituted the mainstay of al-Awda in the area. AQM also killed several insurgent commanders.[16] Funding became another source of conflict between AQM and local Iraqi insurgents. AQM enjoyed significant funds when it first began operations, but over time it began to interfere in the funding operations of local insurgent groups by extorting money from local businesses and traders affiliated with or already paying protection money to local groups. AQM operatives also began robbing travellers of money and valuables and, importantly, encroaching on the lucrative smuggling activities of the sheikhs and local insurgent leaders.[17]

The common ideological foundations of the Iraqi Salafists and the transnationalists of AQM, and their record of joint operations, did not prevent conflict. A 1920 Revolutionary Brigades member with the *nom de guerre* Haj Muhammad Abu Bakr acknowledged in 2007 that the two groups had at one point had a common goal of resisting the occupation, but that now 'we have some disagreements with al-Qaeda, especially about targeting civilians, places of worship, state civilian institutions and services'.[18] Khalaf al-Ayan, a member of the Sunni parliamentary bloc Al Tawafuq, reiterated this point when he stated that all the mainstream insurgent groups had serious differences with AQM because it was pursuing an international and not an Iraqi agenda.[19]

Tensions between mainstream Iraqi insurgents on the one hand and AQM and the ISI on the other were not confined to Anbar province, but spread into Diyala province east of Baghdad, where AQM killed or kidnapped several Sunni insurgent leaders and religious and academic figures. In March 2007, AQM fighters allegedly killed a senior leader in the 1920 Revolution Brigades, who was also the nephew of Harith al-Dari, the most prominent Sunni cleric in Iraq. His murder helped to set off a major verbal assault on al-Qaeda by the 1920 Revolution Brigades. Mahmud al-Zubaydi, a spokesman for the 1920 Revolution Brigades, said: 'Regarding our differences with Al-Qa'ida I prefer to refer this question to them. We had no problem with them. However, many attacks were launched by them in the field, the recent of which was the killing of one of the field commanders, Sheikh Harith, may God have mercy on his soul'.[20] Al-Zubaydi maintained that the Brigades did not target civilians – implying that AQM did – and that his own group was focused on 'fighting against the occupier, its henchmen, lackeys, and agents. It does not attack civilians and has never claimed any such operation'.[21] When asked whether his group agreed with al-Qaeda on operations against the Coalition but disagreed with it on the issue of killing civilians, al-Zubaydi answered: 'Yes, exactly.'[22]

In spring 2007, the IAI launched a sustained propaganda assault on the transnational jihadists, publishing a nine-page letter urging Osama bin Laden to assume religious and organisational responsibility for AQM and suggesting that it was 'not enough to disown [its] actions, it is imperative to correct the path'.[23] In an unprecedented attack, the IAI leadership accused AQM of illegitimate actions including killing 30 IAI members; tarnishing the image of the IAI; engaging in open conflict with other major resistance groups; and robbing from and killing Sunni civilians, including clerics. The group also charged that AQM's methods transgressed Islamic law.[24] The letter did not acknowledge the ISI at all, referring throughout to AQM rather than to the ISI.[25] The IAI also attacked AQM for killing civilians with chlorine gas, implying that this had been an immoral method, and denounced it for killing IAI leaders and clerics who had expressed willingness to negotiate with the Americans over their exit from Iraq. It also expressed disapproval of AQM's tactics of imposing Taliban-type mores and codes of conduct in Sunni areas. A number of individual IAI commanders also made clear their distaste for AQM. One, Abu Mohammad al-Salmani, said in early 2007 that 'Al-Qaeda has killed more Iraqi Sunnis in Anbar province during the past month than the soldiers of the American occupation have killed within three months. People are tired of the torture. We cannot keep silent anymore.'[26] Other insurgent groups were also critical of AQM. At around the same time, an AM commander in Baquba in Diyala province was quoted as saying: 'We do not want to kill the Sunni people nor displace the innocent Shia, and what the al-Qaeda organisation is doing is contradictory to Islam.'[27]

Around six months earlier, in October 2006, stunned by the force of attacks from mainstream insurgents that were already in evidence, the Mujahadeen Shura Council began an intense media campaign. The intent was to establish Abu Hamza al-Muhajir as a successor to al-Zarqawi as head of AQM, and to promote the Mujahadeen Shura Council as a leading force for unity in the defence of Iraq's beleaguered Sunnis. AQM leaders argued that the United States was on the verge of defeat in Iraq, and was spreading lies and rumours in an effort to create divisions among insurgents.[28] But this counter-campaign generated little support or sympathy from Iraqi Salafists or other insurgents. On the contrary, they increased the intensity of their opposition. An Iraqi Salafist with the *nom de guerre* Abu Usama al-Iraqi called on bin Laden to cut his links with AQM because of its 'deviant' behaviour under al-Zarqawi. AQM, al-Iraqi argued, had made a mistake in selecting al-Zarqawi as its leader; AQM should be led by an Iraqi, not by foreigners such as al-Zarqawi and al-Muhajir.

In December 2006, Jihad al-Ansari, founder of the Ali ibn Abi Talib and al-Marsad Brigades insurgent groups, posted an open letter on a popular jihadist website addressed to the ISI and its purported leader, Umar al-Baghdadi. Entitled 'The Solid Structure', the letter questioned AQM's call for allegiance and the appointment of al-Baghdadi as ISI leader without consultation. It was not received well by AQM-affiliated groups. In March 2007, al-Ansari followed up the letter with another, entitled 'After the Solid

Structure', in which he criticised the ISI for ignoring his call for clarification about the intentions behind the ISI's creation. Al-Ansari attacked AQM for taking what he saw as the harmful step of setting up a supposed 'Islamic state' without consulting other groups:

> This step of yours, which you have attempted to secure by every means, has caused innumerable negative results that have harmed the jihadists as a class and hurt the reputation of jihad and mujahadeen. One of the most serious and prominent repercussions of your one-sided step has been that many of your groups and organisation members, desiring to confirm your authority over Muslims in Iraq, have presumed to attack one and all and have continued to incite against anyone who abstains from swearing allegiance to you … We say all this knowing that the organisation [has not always been] the way it is today. However, your headstrong effort to gather individual mujahadeen to your banner, thinking that this would strengthen the power of your organisation and facilitate the independent proclamation of your state, is precisely what has made your organisation adopt horizontal expansion (quantity) at the expense of vertical expansion (quality).[29]

Relations between the mainstream insurgency and AQM/ISI worsened quickly during 2007. In February, Al-Zawraa, a pro-insurgent satellite channel owned by Misha'an al-Jiburi, a former Ba'athist, attacked AQM and the ISI, charging them with extremism, sectarianism, terrorism and arrogance towards other insurgent groups. Al-Jiburi himself openly rejected the ISI's pretensions to leadership of the insurgency, describing AQM as a sectarian group made up of foreigners, whose global agenda had harmed and divided Iraqis. Stung, the ISI lashed out. In March 2007, an audio speech, apparently by al-Baghdadi, alleged that the ISI faced a 'vast conspiracy' hatched by the US and Iraqi governments, Sunni Arab political parties and other insurgents, including jihadist groups. The goals of this conspiracy were to break the supposedly powerful bonds and solidarity between the ISI and Iraqis, to use jihadist groups to strike at the ISI, to distance the global jihadist movement from the battlefield and to eliminate the jihad in Iraq.[30]

This spirited defence of the ISI did not stem the tide of criticism from mainstream insurgents. In April 2007, the IAI launched a major rhetorical attack, supported by Kuwaiti Salafist ideologue Hamid al-Ali. The creation of the ISI, argued the IAI leadership, was a mistake because it did not have the requisite resources to maintain itself and was divisive, since not all groups within the insurgency supported its creation. In May 2007, the ISI released a survey of the insurgency's fourth year, which argued that establishing the ISI had been a critical step towards insurgent unity under a single leadership, namely its own. The survey identified the ISI's broader strategic programme as being the thwarting of Iranian expansionism in Iraq, the prevention of Shia domination and fighting to stop the establishment of a secular client regime. The survey included criticism of Sunni insurgent movements, especially the IAI, for their willingness to negotiate with the Coalition.[31]

The tribes' conflict with AQM

AQM also came into conflict with a significant group of Sunni tribal sheikhs in Anbar and increasingly in other provinces, such as Salahuddin and Diyala, in 2006 and 2007. A suicide bombing in Ramadi in Anbar in January 2006 exposed considerable latent tensions between the jihadist insurgents grouped around al-Zarqawi and the tribes. The attack killed more than 70 Sunni police recruits from local tribes, who had been persuaded to join the security forces by their sheikhs. The sheikhs' rationale for this had been twofold: firstly, work with the security forces would provide income for restless young men; and, secondly, the Sunni community preferred having Sunni security personnel policing their areas than units dominated by Shi'ites. Following the attack, tribal chiefs in and around Ramadi informed AQM that it was no longer welcome, and began a series of meetings to formulate a strategy towards the group. Al-Zarqawi's response was to order the killing of several prominent Sunni clerics and sheikhs.[32] In August 2006, Anbar's tribes took the significant step of forming the Al-Anbar Salvation Council (ASC), under the leadership of Sheikh Sattar Buzaigh al-Rishawi.[33] The council was the clearest indication yet of the tribes' mobilisation against the extremists. Based in Ramadi, the ASC comprised 25 tribes opposed to AQM.[34] The formation of this tribal front, which came to be known by the Americans as the 'Anbar Awakening', provoked a belligerent response from an AQM spokesman: 'We have the right to kill all infidels, like the police and army and all those who support them. This tribal system is un-Islamic. We are proud to kill tribal leaders who are helping the Americans.'[35] In late September 2006, the Mujahadeen Shura Council released a statement threatening several tribes and an insurgent group based in Al-Qaim. It also accused the leaders of the Abu Rughal tribe and the Hamza Battalions, a nationalist insurgent movement consisting mainly of members of the Albu Mahal tribe, of betrayal and treason. The statement argued that it was right to fight this group because it had sided with the enemies of Islam and had violated Islamic law. Its members were now subject to 'traditional punishment', including exile, amputation and crucifixion.[36] In October 2006, the Mujahadeen Shura Council began a media campaign targeting tribal leaders in Anbar province in response to their growing cooperation with the Iraqi government. Abu Hamza al-Muhajir offered the 'traitors' amnesty if they publicly renounced the Iraqi government and accepted AQM's leadership in the insurgency.

What lies behind the rift? Firstly, like other mainstream insurgents, the tribes were fighting because of specific grievances, both material and related to local questions of identity. Most were not interested in pursuing AQM's grandiose goal of a theocracy or caliphate. Secondly, as a result of growing contact with Sunni political parties and individuals, several insurgent groups had become convinced that their community would be politically empowered if it participated in the provincial and national-assembly elections in December 2005 and January 2006. The head of mainstream Sunni Islamist party the Iraqi Islamic Party held talks with insurgent leaders, urging them to suspend their activities during the elections and help persuade the Sunni community to vote. The Islamic extremists were vehemently opposed to the elections. Thirdly, like the mainstream insurgents, the tribes were increasingly repelled by al-Zarqawi's targeting of civilians. In the words of Sheikh Osama al-Jada'an, a prominent tribal leader in Anbar later assassinated by AQM: 'We realized that these foreign terrorists were hiding behind the veil of the noble Iraqi resistance. They claim to be striking at the US occupation, but the reality is they are killing innocent Iraqis in the markets, in mosques, in churches, and in our schools.'[37] Other tribal leaders also spoke up. Sheikh Haqi Ismail al-Fahdawi of the Bu Fahd tribe lamented that his tribe had gone through a 'difficult period' and had lost 'many dear sons'. Al-Fahdawi also expressed deep concern that some young tribal men had been seduced by AQM's message and had joined the organisation. According to another local leader, Hussein Zubeir, 'if it was not for the coyotes among us, no one would have been killed, kidnapped, or bombed'. The chlorine-gas attacks of early 2007 of which Sunnis were casualties outraged the tribes, which viewed them as the desperate acts of an unscrupulous enemy.

The ISI responded to the increasing prominence of the ASC during 2007 with a hostile media campaign. The key message was that the ASC was not as effective as it claimed to be, and that support for it among Anbar's tribes was not extensive. The campaign denounced 'those who sold their religion for a few dollars and were happy to be the crusader occupier's hench-men and a poisoned dagger that stabs the mujahadeen in the back'.[38] The campaign referred to the ASC as the 'Anbar Infidel Council', composed of 'murderers, highway robbers and looters'. The tribes responded with their own information campaign, in which they made the claim that Iran was supporting AQM. ASC member Colonel Fadil Mukhlif al-Dulaimi stated that weapons and explosives from Iran had been discovered at AQM bases in Al-Sufiya, Albu Fahd and Ramadi. The claim was made in the context of a group that traditionally viewed much of its troubles in terms of Iran. To the highly nationalist Sunni Arab tribes, thousands of whose members died in the Iran–Iraq War, Shia Iran is often the root cause of the prob-lems Iraq has faced. In this view, Iran supported the Coalition's invasion and occupation of Iraq, and attempted to use the Iraqi Shia community to subjugate the country's Sunni population, while many of AQM's actions in Iraq seemed similarly to undermine the position of the Sunni community.

Incoherence at the centre of political power

The Sunni community's political incoherence constitutes one of the gravest weaknesses of its resistance movement. Neither the insurgents nor legiti-mate or quasi-legitimate Sunni political parties and organisations have proved capable of playing an effective political role on behalf of the people they purport to represent.

The insurgents' political weaknesses are numerous. Their political incoherence results largely from factionalism and internecine rivalries both between and within groups. This has been reflected in an inability to present a meaningful, united political platform. This failure has been acknowledged by numerous insurgent organisations, and it was with the intention of resolving this profound weakness that many insurgent groups developed political wings and put members in front of the media as spokesmen. The IAI has been one of the more successful in this endea-vour, and the outside world has been able to learn a great deal about its ideology and goals from the regular media appearances made by its key spokesmen. Other groups have followed suit, and some now grant exten-sive interviews to international and Western media.[39] One of the most prominent examples of this is the Islamic Resistance Movement: Hamas-Iraq, a spin-off of the 1920 Revolutionary Brigades. Shortly after it split from its parent group, the head of the group's newly established 'political bureau', Dr Ahmad Abd al-Aziz al-Sa'dun, appeared on Arabic TV station al-Jazeera to lay out the movement's characteristics and broad political goals, saying that the group was 'an independent movement and part of the Islamic-resistance current in the *umma*. It seeks to elevate the *umma* in all the economic, scientific, and economic fields of life and to liberate its will through legitimate means from all the tools of external pressure and hegemony.'[40]

However, political wings that articulate beliefs and goals mean little without traction at the centre, at the level of national politics. Legitimate Sunni political parties and organisations have not performed well, either in representing the Sunni community politically or as ready-made political wings of the insurgency. When the Ba'athist regime fell in 2003, the hitherto dominant Sunni community found itself devoid of well-organised political parties or organisations.[41] In contrast, the Shi'ites and Kurds possessed a large number of well-funded, well-organised parties, many of which had been in exile.

In the absence of effective grass-roots leaders, Sunni clerics pressed for an increased role at the national level, and began to build a political base in Baghdad. The Iraqi Islamic Party, which is affiliated with the Muslim Brotherhood, was among the legitimate Sunni political organisations that reactivated themselves with the fall of the regime. It faced considerable obstacles and problems from the very beginning. Firstly, it was not well organised or well funded. Secondly, as a consequence of the first problem, its network was neither extensive nor deep. Thirdly, it was constantly harassed by Coalition forces, who suspected it of collaborating with Sunni insurgents.[42] Fourthly, its participation in the political process in the immediate post-invasion period meant that it was viewed with distaste by many ordinary Sunnis, and with suspicion by the insurgents. Its ongoing role as the leader of Sunni coalition the Iraqi Accord Front has done little to increase support for it in subsequent years.

One of the earliest efforts to create a coherent Sunni political organisation was undertaken by cleric Harith al-Dari, though his star has since dimmed. Shortly after his return to Iraq from the United Arab Emirates following the ousting of Saddam, al-Dari established the Association of Muslim Scholars (AMS). The AMS argues that Iraqis have a legitimate right to oppose the occupation and to combat it by force. The group refers to the fight against the foreign presence as a *muqawama* (war of resistance), and has called for a united opposition joining Sunnis and Shi'ites against the Coalition. Al-Dari himself sought on numerous occasions to build links with Shia leaders. This became difficult over the course of 2006 as relations between the two communities became increasingly bitter, and Shia-dominated governments in Baghdad accused the AMS of direct support for the insurgency. On this point, al-Dari has said that:

> Our stance on the resistance in Iraq is the same stance as that of every nationalist and Muslim whose land is occupied. If he does not resist, then he supports the resistance. Our sharia orders us to resist the enemies. We do not incite resistance, but we do support it … It is a natural right for any nation.[43]

Al-Dari has steadfastly maintained that resistance is a legitimate right of the Iraqi people, guaranteed by UN resolutions and by international law.[44] For him, the way out of the violence is through the departure of foreign forces, because the occupation uses the strategy of divide and rule to turn Iraq's communities against one another.

Time has not lessened al-Dari's hostility, either to the foreign presence in Iraq or to the prevailing political system. In March 2007, he reiterated his views in an extensive interview with the Beirut-based newspaper *al-Safir*. The political process, he claimed, had failed to liberate Iraq and rescue it from the instability and insecurity the occupation had created. He argued that the political process had not begun the process of reconstruction, was established on sectarian and ethnic foundations and promoted a federalism that was intended to divide Iraq.[45] Al-Dari shares with the mainstream insurgent groups a desire to replace the existing post-Saddam political system with a strong government backed by a strong national army loyal to the Iraqi nation, rather than to factional and sectarian interests.

Al-Dari's insistent recognition of the legitimacy of the resistance and his support for the goals of mainstream insurgent groups could have enabled some of these groups to use the AMS as a platform to represent their views to their opponents. There are several reasons why this did not happen. Firstly, despite the common ground between al-Dari and the AMS and the insurgents, many of the latter were reluctant to see him emerge as the leader of the political wing of the Sunni insurgency. It is not entirely clear why this was the case, though it could stem from internecine rivalries and jealousies. Secondly, al-Dari's self-righteous and inflexible stance towards the Coalition and the Shia-dominated governments in Baghdad preclude the flexibility and pragmatism necessary for negotiations. Thirdly, al-Dari's contempt for the government is reciprocated by its officials, who see him simply as an insurgent. Finally, it is not clear how relevant the AMS remains to Sunni politics. In the early days of the insurgency in 2003–05, when the Sunnis lacked political organisations to represent them, the AMS might have emerged as a serious political player. From 2006 onwards, however, a number of competing groups arose to promote Sunni interests, including the 'Awakening Councils' formed by tribal sheikhs intent on eradicating the jihadist presence. The opportunity has passed for the AMS to play an important role in the political process on behalf of Sunnis.

By late 2007, the Sunni insurgency was in a state of disarray, and strife within the Sunni community was at an all-time high. Many insurgent groups blamed AQM for the decline in the fortunes of the insurgency. Others pointed to the Awakening Councils, accusing them of exacerbating divisions within the Sunni community and deflecting Sunnis from the struggle against the occupation. Some insurgents continued to fight the Coalition, the government and a weakened AQM in their areas. Others drew back from the struggle against US forces, preferring to wait out the 'surge' in the US troop presence that began in early 2007, regrouping and retraining, presumably intent on rejoining the fray once the surge had exhausted itself.[46] The Awakening Councils were also split. Some rejected too close a relationship with Coalition forces.[47] Sheikh Ahmed Abu Risha, the head of the Awakening movement in Anbar since the assassination of his brother Sheikh Sattar Buzaigh al-Rishawi in autumn 2007, has ambitions of entering national politics, and was using his military successes against AQM to begin forming a political movement.[48] Other insurgent groups were simultaneously fighting against and engaging in negotiations with Coalition forces.[49] Some Sunnis joined forces with Shi'ites at local and neighbourhood levels to fight the remnants of AQM.[50] Others chose to form 'fronts' to challenge whichever actors they opposed, whether AQM, the Coalition, the Awakening Councils or the government in Baghdad.[51]

Sanctuaries and external support

Some analysts have argued that support and sanctuary offered by external powers, particularly neighbouring ones, is a significant factor in the success of insurgencies.[52] Conversely, the lack of such support contributes significantly to the possibility of failure. Foreign governments can provide money, weapons and other resources. Sanctuary – the provision of territory for training and housing of insurgent leaders and cadres – is one of the most important resources that foreign powers can provide. The George W. Bush administration's view that the Iraqi insurgency has gained a high level of official support from neighbouring Syria and Iran is not borne out by the evidence. The insurgents have themselves attacked Arab and Islamic governments for their failure to lend extensive support to their 'war of national liberation'. Moreover, the insurgents are wary of regional powers, as well as being dissatisfied with the levels of external support.[53] One insurgent leader, 'Dr Zubeidy', reportedly of Ansar al-Sunna, has said that:

> Our organization depends on stores of weapons left by Saddam
> Husayn, or we buy them from the Iraqi army through merchants.
> We try to avoid links with the Syrians. Either they can sell us out
> at any time if there is heavy pressure on them, or we risk being
> completely controlled by them.[54]

An insurgent leader in the 1920 Revolution Brigades, Abdallah Suleiman Omari, stated that Tehran had offered help to some insurgent groups, but added 'we do not trust Iran'.[55] Omari had pleaded for help from Arab and other governments, but appeared to believe that most regional states were afraid of instability and reluctant to offer support lest doing so provoke the United States: 'We are the only resistance movement in modern history that has received no help or support from any country. The reason is that we are fighting America.'[56] While this overstates the case, it is not far-fetched to argue that most Arab and Muslim states – despite their disapproval of the US occupation of Iraq – have not been willing to antagonise the world's most powerful country by supporting the insurgents. Moreover, the insurgents have not won much sympathy abroad for their particular Ba'athist and Islamist ideologies, which do not resonate widely elsewhere in the Arab world, where the Iraqi resistance is celebrated simply for the fact of its opposition to a foreign presence on Arab soil.

Nevertheless, two neighbouring countries have shown sympathy and support for the insurgency. Syria, a Ba'athist state, has generally been friendly to the Iraqi insurgent cause because of its opposition to US policies in the region and because it has suspected that it was itself America's next target. Damascus lent overt support to certain elements of the insurgency in its early years, but has since reduced its formal, official backing and has discouraged foreign fighters from using its territory for transit purposes.[57] Syria's sympathy for the insurgency waned considerably when extremist Islamists became prominent and the headlines from Iraq became dominated by the rising toll of the sectarian strife. However, Syrian support for the Ba'athist and secular elements of the insurgency has remained consistent, albeit complicated by the division of the Iraqi Ba'ath Party into two competing groups and the reluctance of Iraqis to submit themselves to Syrian oversight, despite their use of Damascus as a host.

It would make little strategic sense for Iran to lend support to the Sunni insurgency, given that Tehran does not want to see the re-establishment of Sunni hegemony in Baghdad. But it is possible that Tehran has provided weapons and arms to Sunni insurgents for tactical reasons, in order to keep the US bogged down in Iraq. Meanwhile, Iran has trained, supported and provided arms to a wide variety of Shia political and military organisations since 2003.[58] One of the leading Shia armed groups, the Badr Brigade, began military training under Iran's Islamic Revolutionary Guard Corps at the Iranian military base of Vahdati in the early 1980s, in preparation for the overthrow of Saddam. Iran also supports Jaish al-Mahdi, despite Moqtada al-Sadr's initial dislike of Iran for what he saw as its interference in Iraqi affairs. The provision of Iranian training for Jaish al-Mahdi – via either Arabic-speaking Iranians or Lebanese Shi'ites associated with Hizbullah – has the potential to transform the militia from a ragtag organisation into a potent player. Iranian support for the Shi'ites has helped to perpetuate the Sunni insurgency, fed the Sunni–Shia conflict and done much to undermine the US mission and goals in Iraq.

There has been extensive non-state support for the insurgency from foreign Arabs and other Muslims. This non-state support has been more important than any backing from regional states, and has had a significant impact on the course of the insurgency, though not necessarily to the insurgents' advantage. The first wave of support from non-Iraqis was motivated and radicalised by vague notions of Arab nationalism and religious sentiment, which encouraged foreigners to make their way to Iraq to fight the foreign occupation. Large numbers of Syrian volunteers with close tribal and cultural links to Iraqis across the border felt duty-bound to fight, though they received no encouragement from their government. Salafist and Islamist feeling in towns and rural areas in eastern Syria, along the border with Iraq, had grown in the 1990s and early 2000s.

The first foreign fighters were poorly trained and ill-equipped. Many entered Iraq prior to the invasion or as the Coalition forces advanced towards Baghdad in March and April 2003. Despite their lack of military experience or training, many fought doggedly, even to the death, in battles between Iraqi irregular forces and the Coalition advancing from the south. After *Operation Iraqi Freedom*, some foreign insurgents went home, outraged by the reluctance among Iraqis to continue the fight and by alleged betrayals of foreign fighters to Coalition forces. Others remained and continued fighting. Often, there were more foreign volunteers than weapons available. The fighters suffered from a lack of organisation, attacking targets opportunistically and operating in small groups with little or no command structure, relying on runners and messengers to communicate among themselves. Iraqi insurgent organisations found them more of a hindrance than a help.

The second group of foreign fighters, which entered Iraq from 2004 following the intensification of the insurgency, came from Muslim immigrant communities in Western Europe as well as from Arab countries. Most were motivated by sympathy for Salafist ideology. Many were initially as unfamiliar with weapons and military methods as the first wave that had fought during the conventional phase of the war. They made up for their lack of military skills with their religious motivation and willingness to die for their cause, and these foreign Salafists were heavily involved in suicide bombings from 2004 onwards. As early as 2004, their targeting of civilians had set the stage for a growing Iraqi dislike of foreign Arabs, a sentiment which was to grow dramatically once these fighters came under al-Zarqawi's control.

Over time, the infiltration of foreign Salafists into Iraq became a well-organised endeavour, and foreign fighters in Iraq became more ideologically aware and cohesive, better organised, better trained and more skilled in combat. Nonetheless, many still remained destined for one type of operation: the suicide mission. Many Iraqi insurgent groups preferred to use foreigners in this manner.

Online jihadist magazines provide instructions on how to enter Iraq via neighbouring countries, and on physical and mental preparation for battle. They also provide tactical information, such as lessons learned by fighters in Iraq and information on US standard operating procedures and how to assemble bombs.[59] Jihadist websites also issue daily reports giving the latest intelligence on raids by US-led forces on border points, and on new crossing locations.[60]

The structural weaknesses of the Sunni insurgency account for much of its manifest failure as a movement of national liberation. Moreover, by 2007 Iraqi Sunni insurgents were fighting a former tactical ally, AQM, in a vicious internecine battle in which Iraqi Sunnis sought the aid of their erstwhile enemies, the Americans. The near-collapse of AQM in 2008 – partly at the hands of Sunni and US forces – focused the attention of the Sunni community on its future in the new, Shia-dominated Iraq. But there has been a subtle and profound change in the mindset of many former insurgent groups and leaders; they are no longer fighting to restore the past. They are now thinking of fighting to get what they can in the new Iraq. They wish to maximise their gains and to minimise Shia domination of Sunni-majority areas. Whether they will work for these aims by political or by violent means remains an open question, and is dependent on how the government deals with them.

CONCLUSION

The overthrow of Saddam Hussein in April 2003 was supposed to lead to the emergence of a new Iraq. It did, but not in the way policymakers in Washington had envisaged. Instead of a secure, stable and transformed Iraq moving towards democracy and the rule of law, the country descended into violence.[1]

The war between the Coalition and the outmatched Iraqi military ended on 9 April 2003, the day Saddam's statue was toppled in Baghdad's Firdos Square. Looting began in the major cities, before the violence turned into a Sunni-inspired rejection of the occupation.[2] The insurgency began as an ad hoc and disorganised resistance by disgruntled individuals and groups, primarily in Sunni areas such as Falluja and Tikrit, where hostility to the US presence was intense.[3] But by late 2003, the influx of disaffected former Iraqi military personnel following the disbanding of the Iraqi Army had ensured that the insurgency would become more dangerous. In 2004, Shia elements under the leadership of Moqtada al-Sadr joined in the fighting against the Coalition. Coalition forces were hard-pressed to deal with what appeared to be becoming a national-liberation struggle. But there was little if any coordination between Sunni and Shia insurgents. An undeclared civil war between the recently empowered Shia majority, which now controlled the government, and the marginalised but formerly dominant Sunni minority began in 2005, and was the dominant theme in 2006.[4] Many observers concluded that Iraq's future was grim.[5]

Even as Sunnis and Shi'ites fought each other, there were also signs of a split within the Sunni insurgency, between local insurgents and the tribes on one side and the Salafi-jihadist strain on the other.[6] Tensions between the two groups – one fighting to rid Iraq of foreign occupation and restore Sunni paramountcy, the other fighting to establish a purist Islamist political system – had emerged as early as 2004, in the insurgent bastion of Falluja, but it was not until 2006 that the course of the insurgency was affected. Many of the local insurgents, including some of the most Islamist among them, were antagonised by AQM's tactics and disagreed with its goals.[7] The Sunni tribes in particular bore the brunt of AQM's disruptive strategies in the rural hinterlands.[8] The Islamist extremists imposed themselves on the tribes in a manner they dared not do in relation to the heavily armed Sunni insurgents, made up as they were of former military and security personnel and local Islamists. The AQM leadership apparently believed that the tribes could be easily manipulated and used as a resource in the struggle against the Coalition and the Shi'ites.

Violence in Iraq reached a peak in 2006. But it was structurally complex, with multiple layers to the fighting. A Sunni insurgency against the Coalition was taking place concurrently with violence within the Sunni community. Simultaneously, Sunni and Shia Arabs were fighting a brutal and undeclared civil war against each other. The intensity and brutality of this conflict increased dramatically after the February 2006 bombing of the al-Askariyya shrine in Samarra. The Shi'ites themselves were divided by class, between al-Sadr's movement, which represented the poor, and the middle-class and intellectuals represented by the Islamic Supreme Council of Iraq and the al-Da'wa party.[9] Tensions were also high between Arabs and Kurds in flashpoints such as the cities of Kirkuk and Mosul, and in provinces such as Diyala.[10] Furthermore, there was little or no movement at the national political level towards resolving pressing issues such as reconciliation between the communities, the integration of the Sunnis and the equitable sharing of oil revenues.

The mood in Washington was bleak in late 2006. There was a strong sense within the administration that the US was heading for a monumental defeat in Iraq, and domestic pressure for the US to do something decisive, such as setting a definite timetable for withdrawal. On the advice of retired officers and civilian analysts, however, the Bush administration decided on a 'surge' in the US troop presence, to reduce the violence and establish the conditions in which Iraqis could bring about national reconciliation. The surge had a dramatic impact, lessening the violence by the end of 2007 and leading to claims that the US was on the verge of victory. The noticeable signs of improvement in the country were testament to the ability of the US to learn and implement effective counter-insurgency measures on the one hand, and to the structural problems, failures and mistakes of the Sunni insurgency on the other. The gains are, however, tenuous, and they could be reversed.

In 2008, there was less talk of imminent victory, as even fervent believers in the surge came to recognise the full complexity of the situation in Iraq. Security remained fragile. General David Petraeus, commander of the Multi-National Force in Iraq, driving force behind the revamping of US counter-insurgency doctrine and implementer of the 'surge', sounded a note of caution as his tour of duty ended in mid September 2008. His replacement, General Raymond Odierno, was equally cautious. Both officers correctly emphasised that the gains from the surge were fragile and potentially reversible. But the issue is more deep-seated than this: the success of the surge may be not only transitory but also illusory, in that many of the structural conditions for a diminution of the violence in Iraq either existed before the implementation of the surge or appeared in parallel with, but were not due to, it.[11] The homogenisation of neighbourhoods by 2007 meant that there was now less opportunity for Sunnis and Shi'ites to kill each other.[12] Al-Sadr's August 2007 decision to demobilise his militia, which had been heavily engaged in fighting Sunnis and provoking US forces, was another significant factor in reducing the violence. Al-Sadr and his senior commanders then identified and eliminated those alleged to be rogue elements within Jaish al-Mahdi.

The most important structural factor, though, has been the dramatic change in the Sunni insurgency with the creation of the Awakening Councils in Anbar. The successes of these tribal councils contributed to the rise of similar militia in Baghdad and other provinces, such as Salahuddin and Diyala, where the inhabitants and local insurgents had finally had enough of the depredations of the extremists.[13] The Awakening in Baghdad, particularly in the heavily Sunni neighbourhoods of Ghazaliya and Amiriya, was not tribal, but was composed of insurgents from the former regime's bureaucracy, civil service and intelligence and military services. They took the initiative in organising the struggle against the extremists, but it was not long before the US military recognised the potential benefits of providing support in the form of arms and money.[14] Thus, the US helped to create a major armed force outside the control of the central state.

As the former insurgents celebrated their victories against AQM, promoted their alliance with the Americans, implicitly showing the Shia-dominated government that they had strong backing from the most powerful force in the country, and expanded their militia organisations, tensions emerged between them and the government of Nuri al-Maliki. Neither side had high regard for the other.[15] Sunnis viewed with dismay the entrenchment of Shi'ites in the Iraqi body politic. Some were still fanatically dedicated to ending this domination, while others were more pragmatic, demanding resources for reconstruction and the absorption of personnel into the security services. Having helped the US to defeat a common enemy in AQM, the Sunni militias are now asking themselves whether they should once again take on their second-largest enemy, the Shi'ites. Now, however, Shi'ites dominate the state and are increasingly self-confident, as evidenced by their muscular politics, both towards the Americans concerning the reduction of the US military presence, and vis-à-vis the Sadrists within the Shia community. The US has grown increasingly uncomfortable with al-Maliki's assertion of power, but has put a brave face on it, claiming that it shows Iraqi self-assertiveness and confidence in Iraq's own capabilities. As Iraq rather grimly celebrated the fifth anniversary of the fall of Saddam's regime, violence broke out between the government and Moqtada al-Sadr's Sadrist movement. Divisions between the government and this large and disaffected Shia movement continued despite a much-touted government offensive against Jaish al-Mahdi in Basra. By May 2008, the government had gained the upper hand, as better-prepared Iraqi forces poured into Sadr City in Baghdad and consolidated control over Basra.

Given its increased self-confidence, al-Maliki's government is in no mood to bargain with the Sunni community. It does not feel the need to 'reward' the Sunni militias, whom it sees as former insurgents with blood on their hands. The government also fears that a substantial re-entry of Sunnis into the armed forces could threaten the hard-won Shia dominance of the military. The existence of a large group of armed former Sunni insurgents led by men with political ambitions worries the government in Baghdad. Their military strength would give these Sunni groups considerable weight should they execute an entry into national politics. The government is prepared to tolerate Shia militias as long as they do not overstep their bounds. It is also prepared to tolerate the Kurdish *peshmerga*, since they do not tend to get involved in the Iraqi political process.[16] But the Sunnis are a different matter, and the deep-seated hostility between the two communities may prompt renewed violence. The government's stubbornness stems in part from the perception – which is probably accurate – that the balance of power in the country has turned in favour of the Shia majority, and that it does not need to seek favour with the armed Sunni

groups. This assessment is dangerous: although the Shi'ites are likely ultimately to secure their dominance – particularly in view of the build-up
of the government's armed forces and support from Iran – the possible
renewal of hostilities as the US gradually reduces its presence could well
result in more severe strife than the 'undeclared civil war' of 2005–06.

NOTES

Introduction

1 For more details, see Bard O'Neill, *Insurgency and Terrorism: From Revolution to Apocalypse*, 2nd edition (Dulles, VA: Potomac Books, 2005), pp. 1–70; Klaus Knorr, 'Unconventional Warfare: Strategy and Tactics in Internal Political Strife', *Annals of the American Academy of Political and Social Sciences*, vol. 341, May 1962, pp. 53–64; Chalmers Johnson, 'Civilian Loyalties and Guerrilla Conflict', *World Politics*, vol. 14, no. 4, July 1962, pp. 646–61.

2 For an extensive analysis of the nature and characteristics of internal warfare, see Harry Eckstein, 'On the Etiology of Internal Wars', *History and Theory*, vol. 4, no. 2, 1965, pp. 133–63. This study uses the terms 'insurgency', 'internal war' and 'guerrilla war' interchangeably.

3 Captain Peter Layton, 'The New Arab Way of War', *Proceedings*, March 2003, http://www.military.com?Content/ MoreContent1?file=NI_New_0303; Andrew Bacevich, 'The Islamic Way of War', *American Conservative*, 11 September 2006, http://www.amconmag. com/2006/2006_09-11/print/coverprint. html; Sydney Freedberg, 'Iraqi Rebels: The New Iraqi Way of War', *National Journal*, 9 June 2007; Ron Laurenzo, 'Middle East Foes Seek to Blunt US Edge', *Defense Week*, 14 June 1999, p. 7.

4 The best-known and probably the most learned jihadi strategic thinker is Syrian Abu Mus'ab al-Suri (aka Mustafa Setmariam Nassar), who was captured in Pakistan in November 2005. Al-Suri is the author of a massive tome of almost 1,600 pages called *The Global Islamic Resistance Call*, which analyses the best resistance strategies for Islamist movements. For a summary of his views, see Andrew Black, 'Al-Suri's Adaptation of Fourth Generation Warfare Doctrine', *Terrorism Monitor*, vol. 4, no. 18, 21 September 2006, http://www.jamestown.org/terrorism/ news/article.php?articleid=2370137.

5 On the Jewish insurgencies against Rome see Josephus, *The Jewish Wars* (Harmondsworth: Penguin Books, 1978).

6 Al-Hajjaj bin Yusuf al-Thaqafi was a well-known teacher, renowned soldier and solid and ruthless administrator. His bloodthirsty speech delivered to the inhabitants of Kufa is widely known throughout the Arab world and can be recited by heart by many Iraqis. Kufa was founded as a garrison town for soldiers from Arabia following the capture of Mesopotamia from the Persian Empire; for more details and the quote see Abdel Wahid Dhanun Taha, *Al-Iraq fi ahd al-Hajjaj bin Yusuf al-Thaqafi: Min al nahya al-siyasiyah w al idariyah* [Iraq During the

Era of al-Hajjaj bin Yusuf al-Thaqafi: From a Political and Administrative Perspective] (Benghazi: Dar al-Maktab al-Watani, 2004).

7   Albert H. Hourani, *A History of the Arab Peoples* (Cambridge: Belknap Press, 1991), p. 251.

8   For the Mameluke power structure and relations with the tribes in Iraq, see Tom Nieuwenhuis, *Politics and Society in Early Modern Iraq* (The Hague: Martinus Nijhoff Publishers, 1982).

9   For details of Ottoman state-building efforts and the obstacles they faced, see Soli Shahvar, 'Tribes and Telegraphs in Lower Iraq: The Muntafiq and the Baghdad–Basra Telegraph Line of 1863–65', *Middle Eastern Studies*, vol. 39, no. 1, January 2003, pp. 89–116; Charles Tripp, *A History of Iraq* (Cambridge: Cambridge University Press, 2007), pp. 8–29.

10 Stuart Cohen, 'Mesopotamia in British Strategy, 1903–1914', *International Journal of Middle Eastern Studies*, vol. 9, no. 2, April 1978, pp. 171–81.

11 For details on the British debates see Peter Sluglett, *Britain in Iraq: Contriving King and Country* (London: I.B. Tauris, 2007), pp. 8–41; Toby Dodge, *Inventing Iraq: The Failure of Nation Building and a History Denied* (New York: Columbia University Press, 2005), pp. 5–42; V.H. Rothwell, 'Mesopotamia in British War Aims, 1914–1918', *Historical Journal*, vol. 13, no. 2, June 1970, pp. 273–94.

12 On the causes, evolution and tactics of the Iraqi Revolt of 1920, see Ghassan Attiyah, *Iraq: Nashaat al-Dawla 1908–1921* (Surbiton: LAAM Limited, 1988), pp. 397–456; Phebe Marr, *The Modern History of Iraq* (Boulder, CO: Westview Press, 1985), pp. 32–34; Tripp, *A History of Iraq*, pp. 39–44; Kristian Ulrichsen, 'The British Occupation of Mesopotamia, 1914–1922',

Journal of Strategic Studies, vol. 30, no. 2, April 2007, pp. 349–77; Mark Jacobsen, 'Only by the Sword: British Counter-Insurgency in Iraq, 1920', *Small Wars and Insurgencies*, vol. 2, no. 2, August 1991, pp. 323–63; Graeme Sligo, 'The British and the Making of Modern Iraq', *Australian Army Journal*, vol. 2, no. 2, Autumn 2005, pp. 275–86.

13 For details of counter-insurgency through 'air control' see Priya Satiya, 'The Defense of Inhumanity: Air Control and the British Idea of Arabia', *American Historical Review*, vol. 111, no. 1, http://www.historycooperative.org/journals/ahr/111.1/satia.html.

14 James Wagner, 'Iraq', in Richard Gabriel (ed.), *Fighting Armies* (Westport, CT: Praeger Publishers, 1987), p. 63.

15 David Baran, 'L'adversaire irakien', *Politique Etrangère*, no. 1, 2003, pp. 59–75.

16 Joby Warrick, 'Uncertain Ability to Deliver a Blow', *Washington Post*, 5 September 2002, p. 1; Molly Moore, 'A Foe That Collapsed from Within', *Washington Post*, 20 July 2003, p. 1. On Saddam Hussein's decision-making style, see Kevin Woods, James Lacey and Williamson Murray, 'Saddam's Delusions: The View From the Inside', *Foreign Affairs*, May–June 2006. For more theoretical detail on the creation of effective conventional military power, see Risa Brooks, 'The Impact of Culture, Society, Institutions and International Forces on Military Effectiveness', in Risa Brooks and Elizabeth Stanley (eds), *Creating Military Power: The Sources of Military Effectiveness* (Stanford, CA: Stanford University Press, 2007), pp. 1–26.

17 Tony Perry and Tyler Marshall, 'Hussein's Irregulars Slow US Advance', *Los Angeles Times*, 28 March 2003, http://ebird.dtic.mil/Mar2003/s20030328168869.html.

Chapter One

1 The modern 'ethnisation' of the conflict between Sunni and Shia Arabs can be said to have begun in 1969–71, when the new Ba'athist regime began to deport Iraqi Shi'ites, whom they classified as Iranians. Following the fall from power of the Sunnis, the notion that the Shi'ites of Iraq were somehow less Arab by virtue of being Shia began to spread – a curious perception, since Iraq was the birthplace of Shiism. The fact that many Shi'ites have returned to Iraq after decades of living in Iran – either because they were exiled there or because they voluntarily emigrated as a result of opposition to the Ba'athist regime – has prompted many Sunnis to take the view that the returnees are more Iranian than Iraqi, and that they are a 'fifth column' – *tabour khamis* – for Iran. For more details on this issue, see Pierre-Jean Luizard, 'Iraniens d'Irak, direction religieuse chiite et Etat arabe Sunnite', *Cemoti*, no. 22 – Arabes et Iraniens, 4 March 2005, http://cemoti.revues.org/document139.html; Ali Babakhan, 'Des Irakiens en Iran depuis la révolution islamique', *Cemoti*, no. 22 – Arabes et Iraniens, 4 March 2005, http://cemoti.revues.org/document140.html; and the lengthy study by Ferhad Ibrahim, *Konfessionalismus und Politik in der arabischen Welt: Die Schiiten im Irak* (Münster: Lit Verlag, 1997).

2 In *Insurgency and Terrorism*, O'Neill refers to a type of insurgency that he calls 'preservationist'. My category of restorationist is similar but more nuanced, since the Sunnis no longer have power to preserve; they are out of power, and began their insurgency as a way of *restoring* their former position.

3 Interview with the author.

4 Quoted in Andrew Marshall, 'Leave Iraq, Tribesmen and Sacked Troops Tell US', *Washingtonpost.com*, 2 June 2003, http://www.washingtonpost.com/ac2/wp-dyn/A2879-2003Jun2?language=printer&content.

5 Brian Conley and Muhammad Zaher, 'A Word from the Islamic Army', Inter Press Service News Agency, 16 May 2006, http://ipsnews.net/news.asp?idnews=33246.

6 Ghaith Abdul-Ahad, 'The US is Behaving as if Every Sunni is a Terrorist', *Guardian*, 26 January 2005, http://www.guardian.co.uk/print/0,3858,5111992-103680,00.html.

7 Conley and Zaher, 'A Word From the Islamic Army'.

8 Juan Cole, *Sacred Space and Holy War: The Politics, Culture and History of Shi'ite Islam* (London: I.B. Tauris, 2002), pp. 16–30; Selim Deringil, 'The Struggle Against Shiism in Hamidian Iraq: A Study in Ottoman Counter-Propaganda', *Die Welt des Islams*, New Series, no. 1–4, 1990, pp. 45–62.

9 For the conversion of the Bedouin Arabs in the south to Shiism, see Yitzhak Nakash, *The Shi'is of Iraq* (Princeton, NJ: Princeton University Press, 2003).

10 Pierre-Jean Luizard and Joe Stork, 'The Iraqi Question From the Inside', *Middle East Report*, no. 193, March–April 1995, pp. 18–22.

11 Foreign Broadcast Information Service (FBIS) Report, Open Source Center, GMP20050519514006, 19 May 2005. The *shu'ubiya* was an intellectual and literary movement during the time of the Abbasid dynasty, launched largely by Iranians. Its purpose was to elevate the non-Arab subject populations of the empire to a level equal to that of the dominant Arabs.

12 Gareth Stansfield, 'The Reshaping of Sunni Politics in Iraq', Al Jazeera Special Reports, 18 March 2004, http://english.aljazeera.net/NR/exeres/56F9DA69-4253-CC9-8F49-DB7EC687E74B.htm; Robert Worth, 'For Some in Iraq's Sunni Minority, a Growing Sense of Alienation', *New York Times*, 8 May 2005, http://ebird.afis.osd.mil/ebfiles/e20050508367202.html.

13 Sabrina Tavernise, 'As Iraqi Shiites Police Sunnis, Rough Justice Feeds Bitterness', *New York Times*, 6 February 2006, p. 1.

14 See p. 19 for an explanation of this term.

15 'Dawlat al Iraq al-Islamiya: Al-taqrir al-akhbari li wilayat Baghdad' [Islamic State of Iraq: News Report on Baghdad Province], Mohajroon.com, 28 November 2006.

16 'Dawlat al-Iraq al-Islamiya: Hawla al-tasa'id al-safawi dhid al-sunna fi Baghdad' [Islamic State of Iraq: About the Latest Safavid Escalation Against the Sunnis of Baghdad], Tajdeed.org. uk, 28 November 2006, http://www. tajdeed.org/uk/forums/showthread. php?threadid=47193.

17 'The Sectarian Battle for Baghdad, Part I: The War of the Neighborhoods', *Global Issues Report*, 22 January 2007.

18 David Baran, 'Iraq: The Party in Power', *Le Monde Diplomatique*, December 2002, http://mondediplo.com/2002/12/05iraq.

19 '"Hamas of Iraq" Seeks to Define its Political and Military Roles', *Global Issues Report*, 3 May 2007.

20 On Jaish al-Rashidin, see an extensive and detailed interview with its leader in 'Jihadist Websites: OSC Summary in Arabic', Open Source Center, GMP20070824301001, 8–13 August 2007, https://www.opensource.gov/portal/ server.pt/gateway/PTARGS_0_0_3386 _212_0_43/http.

21 'Al-Furqan Army Issues First Statement, Declares Separation from Islamic Army in Iraq', 'Jihadist Websites: OSC Summary in Arabic', Open Source Center, GMP20070823050002, 17 August 2007, https://www.opensource. gov/portal/server.pt/gateway/ PTARGS_0_0_3386_212_0_43/http.

22 The following discussion of the role of the tribes in state formation and the Iraqi insurgency, and their relations with AQM and the Americans, is drawn from observations in Iraq in November 2003–April 2004, July–September 2005 and October–November 2007, and from the following academic and journalistic accounts: Amatzia Baram, 'Neo-Tribalism in Iraq: Saddam Hussein's Tribal Policies 1991–96', *International Journal of Middle Eastern Studies*, vol. 29, no. 1, February 1997, pp. 1–31; Stephen Glain, 'Stronghold Can Backfire: Iraqi Tribes are Key Source of Loyalty, Rebellion', *Wall Street Journal*, 23 May 2000; Hugh Pope and Bill Spindle, 'Tribes May Play Crucial Role in Political Future of Iraq', *Wall Street Journal*, 16 December 2002; David Kilcullen, 'Anatomy of a Tribal Revolt', *Small Wars Journal*, August 2007, http:// smallwarsjournal.com/blog/2007/08/ print/anatomy-of-a-tribal-revolt; Falah Jabar, 'Sheikhs and Ideologues: Deconstruction and Reconstruction of Tribes under Patrimonial Totalitarianism in Iraq, 1968–98', in Falah Jabar and Hosham Dawood (eds), *Tribes and Power: Nationalism and Ethnicity in the Middle East* (London: Saqi Books, 2003); William McCallister, 'The Iraqi Insurgency: Anatomy of a Tribal Rebellion', *Firstmonday*, vol. 10, no. 3, March 2005, http://firstmonday.org/issues/issue10_3/ mac/index.html.

23 For detailed analysis see Baram, 'Neo-Tribalism in Iraq'.

24 Joe Klein, 'The Next War in Iraq', *Time*, 22 August 2007, http://www.time.com/time/ nation/article/0,8599,1655219,00.html.

Chapter Two

1   Another example of a highly factionalised insurgency is the mujahadeen insurgency against the Soviet occupation of Afghanistan in the 1980s.

2   In his essay 'Theory of the Partisan', which is only now becoming known in the English-speaking world, Carl Schmitt distinguishes between partisans who fight for the defence of their territory and for parochial or local goals – the 'autochthonous' defenders of the homeland – and world revolutionaries, who are deterritorialised and seek to promote revolutionary goals everywhere. See Carl Schmitt, *Theory of the Partisan* (New York: Telos Press Publishing, 2007), pp. 14–22. In the same way, we can distinguish between 'mainstream' Iraqi groups and the transnational revolutionaries of AQM. This deep ideological fissure was to be of profound significance from late 2006, as the Iraqi groups began to distance themselves from the transnationalists.

3   John Devlin, 'The Ba'thist Party: Rise and Metamorphosis', *American Historical Review*, vol. 96, no. 5, December 1991, pp. 1,396–1,407.

4   Cited in Z.L. Kaul, *Iraqi Revolution in the Service of Humanity* (New Delhi: New Wave Printing, n.d.), pp. 51–64.

5   See 'Izzat Ibrahim al-Duri Interview with *Time*', *Time*, 24 July 2006, http:saddamhusseinblog.blogspot. com/2007/06/izzat-ibrahim-al-duri-interview-with.html; and Bobby Ghosh, 'Inside the Mind of Saddam's Chief Insurgent', *Time*, 24 July 2006, http://www.time.com/time/world/article/0,8599,1218388,00.html.

6   Interviews and observations by the author in Baghdad, Ramadi, Mosul and Tel Afar, November 2003–April 2004 and July–October 2005.

7   Abd al-Wahhab al-Qaysi, 'An Officer in the Resistance Tells *al-Riyad* that Iraqi Resistance to the Americans Will Escalate During the Month of Ramadan', *al-Riyad*, 14 October 2003, in FBIS Report, Open Source Center, GMP20031014000027, 14 October 2003.

8   Patrick Graham, 'Beyond Fallujah: A Year with the Iraqi Resistance', *Harper's Magazine*, June 2004, p. 41.

9   Matthew Campbell, 'Resistance Chief Warns US War is About to Get Bloodier', *Sunday Times*, 21 December 2003, http://ebird.afis.osd.mil/ebfiles/e20031222243765.html.

10 *Ibid.*

11 Ghaith Abdul-Ahad, 'Inside Iraq's Hidden War', *Guardian*, 20 May 2006, p. 4.

12 *Ibid.*

13 The statistics cited in this paragraph are from 'An Overview Submitted by the UN System to the Security Council Panel on Humanitarian Issues', 24 March 1999, CASI Web version prepared 1 October 1999, http://www.casi.org.uk/info/undocs/spec-top.html. This report provides an extensive survey of Iraq's socioeconomic collapse in the 1990s.

14 John Battersby, 'Saddam, Sanctions Plunge Iraq into "Irreversible" Ruin', *Christian Science Monitor*, 30 October 1995, p. 1.

15 Isabelle Lassere, 'Saddam Imposes a Forced March of Islamization', *Le Figaro*, 23 October 2002; Anthony Shadid, 'Iraq's Rising Forces of Faith Create Fears for Future', *Washington Post*, 15 March 2003, p. A15.

16 *Ibid.*

17 Dan Murphy, 'Radical Islam Grows Among Iraq's Sunnis', *Christian Science Monitor*, 28 July 2004, http://www.csmonitor.com/2004/0728/p01s04-woiq.html.

18 For the political fortunes of one such cleric, Abdallah al-Janabi, during Saddam's regime, see Arnold Hottinger, 'Die befreite Stadt des Sunniten predigers Janabi', *Neue Zürcher Zeitung*, 15 September 2004, http://www.nzz.ch/dossiers/2002/irak/2004.09.15-al-article9UVGR.html.

19 Murphy, 'Radical Islam Grows Among Iraq's Sunnis'.

20   *Ibid.*
21   Sophie Shihab, 'Resurgence of Salafist Current in Sunni Crucible Reflects Radicalization Against Americans', *Le Monde*, 16 November 2003, in FBIS Report, Open Source Center, EUP20031118000024, 16 November 2003.
22   Didier François, 'The Salafists Lie in Ambush in Baghdad', *Libération*, 13 January 2004, in Open Source Center, EUP20040113000014.
23   *Ibid.*
24   *Ibid.*; see also Shihab, 'La résurgence du courant salafiste dans le creuset Sunnite témoigne d'une radicalisation contre les américains', *Le Monde*, 15 November 2003, http://www.lemonde.fr/web/imprimer_article/0,1-0@2-3218,36-342023,0.html; Pierre Barbancey, 'Irak: Parmi les imams Sunnites de la résistance', *L'Humanité*, 11 November 2003, http://www.humanite.fr/popup_print.php3?id_article=382335.
25   *Ibid.*
26   Abdul-Ahad, '"The Jihad is Now Against the Shias, Not the Americans"',

Guardian, 13 January 2007, http://www.guardian.co.uk/world/2007/jan/13/iraq.iraqtimeline.
27   'Jihadist Websites', FBIS Report, Open Source Center, GMP20051115371016, 15 November 2005.
28   'The Mujahidin Army in Iraq Issues Statement, Video, Urges US Citizens to Overthrow Administration', 'Jihadist Websites', FBIS Report, Open Source Center, GMP20050321000291, 21 March 2005.
29   *Ibid.*
30   'Jihadist Websites', FBIS Report, Open Source Center, GMP20050324000125, 24 March 2005.
31   See 'Iraq's Civil War, the Sadrists and the Surge', Middle East Report No. 72, International Crisis Group, 7 February 2008.
32   'Jihadist Websites: OSC Summary', Open Source Center, FEA20070708220946, 8 July 2007.
33   *Ibid.*
34   *Ibid.*
35   *Ibid.*

Chapter Three

1   'Jihadist Websites', Open Source Center, FEA20061005028423, 5 October 2006.
2   Pierre Perrin, 'Interview with Husayn al-Falluji: Resistance in Iraq Primarily Islamic', *Libération*, 12 March 2007, in Open Source Center, EUP20070312029008.
3   *Ibid.*
4   'Izzat Ibrahim al-Duri Interview with *Time*'.
5   'Iraq', FBIS Analysis, Open Source Center, GMP20051109388001, 9 November 2005.
6   'Jihadist Websites', FBIS Report, Open Source Center, GMP20050830336001, 30 August 2005.
7   Lydia Khalil, 'Islamic Army in Iraq

Pursues Strategy of Negotiation and Violence', *Terrorism Focus*, vol. 3, no. 41, 24 October 2006.
8   'Jihadist Websites', Open Source Center, GMP20070124302003, 18 January 2007.
9   'Al-Haqq Agency Interviews Political Official of 1920 Revolution Brigades', Open Source Center, GMP200706 14410001, 9 June 2007.
10 'Jihadist Websites', FBIS Report, Open Source Center, GMP20050313000163, 3 March 2005.
11 'Jihadist Websites', FBIS Report, Open Source Center, GMP20070515338001, 14 May 2007.

Chapter Four

1   Author interviews in Baghdad and author observations of insurgent groups, November 2003–April 2004 and July–September 2005.

2   'Islamic Army in Iraq Opposes US, Iranian Occupation; Offers US Conditional Negotiations', 'Jihadist Websites', Open Source Center (OSC) Report, GMP20061109281002, 9 November 2006.

3   'Al-Fatihin Army Hostile to US, Iran, Shiite Militias', 'Jihadist Websites', OSC Report, Open Source Center, GMP20061128281001, 28 November 2006.

4   Jim Michaels, 'Attacks Rise on Security Convoys', USA Today, 9 July 2007, p. 1.

5   Joseph Galloway, 'Bolder Insurgent Tactics Unleashed in Iraq', thestate.com, 24 April 2005; Ellen Knickmeyer, 'Zarqawi Said to be Behind Iraq Raid', Washington Post, 5 April 2005, p. A1.

6   Richard Boudreaux and John Johnson, 'Anatomy of a Rebel Strike', Los Angeles Times, 22 March 2006.

7   Joshua Partlow, 'Troops in Diyala Face a Skilled, Flexible Foe', Washington Post, 22 April 2007, p. 1.

8   Peter Eisler, 'Insurgents Adapting Faster to US Defenses', USA Today, 16 July 2007, p. 1.

9   The Iraq Survey Group was the organisation tasked by the Coalition with investigating Iraq's weapons programmes.

10   The most extensive analyses of the insurgents' first chemical-warfare efforts are in Iraq Survey Group, 'Iraq Survey Group Final Report: Volume III: Iraq's Chemical Warfare Program', Annex E, September 2004, http://www.globalsecurity.org/wmd/library/report/2004/isg-final-report/isg-final-report_vol3_cw-anx-e.htm; Bob Dorgin, 'Report: Iraq Insurgents Seeking Chemical, Germ Weapons', Seattle Times, 10 October 2004, http://seattletimes.nwsource.com/cgi-bin/PrintStory.pl?document_id=2002058978&zsection_id=268448413&slug=wmd10&date=20041010.

11   Alissa Rubin, 'Chlorine Gas Attack by Truck Bomber Kills up to 30 in Iraq', International Herald Tribune, 7 April 2007, http://www.iht.com/articles/2007/04/07/africa/web-0407-iraq-23092.php.

12   Ba'ath Party communiqué, May 2004, quoted in Mike Whitney, 'The Guerilla War on Iraqi Oil', Dissident Voice, 3 January 2006, http://www.dissidentvoice.org/Jan06/Whitney03.htm.

13   Justin Blum, 'Terrorists Have Oil Industry in Cross Hairs', Washington Post, 27 September 2004, p. A12.

14   'Bin Laden Tape Urges Oil Attack', BBC News, 16 December 2004, http://news.bbc.co.uk/1/hi/world/middle_east/4101021.stm.

15   Quoted in Fred Burton, 'Attacks on Energy Infrastructure: Desire, Capability and Vulnerability', Stratfor, 2 March 2006, http://www.stratfor.com/attacks_energy_infrastructure_desire_capability_and_vulnerability.

16   OSC Report, Open Source Center, GMF20060808281005, 8 August 2006.

Chapter Five

1   The noted Iraqi political scientist Wamid Nadhmi, a professor at Baghdad University and an expert on the Iraqi revolt of 1920, seemed to believe that the uprising by the Sadrists in 2004 represented a turning point: 'What is striking is how much has changed in a week. No one can talk about the Sunni Triangle anymore [a majority-Sunni area where most of the violence allegedly took place]. No one can seriously talk about Sunni–Shia fragmentation or civil war.

The occupation cannot talk about small bands of resistance. Now it is a popular rebellion and it has spread'. Wamid Nadhmi, quoted in Rami el-Amine, 'The Shia Rise Up', *Dissident Voice*, 18 May 2004, http://www.dissidentvoice.org/May2004/El-Amine0518.htm.

2 Ahmed S. Hashim, *Insurgency and Counterinsurgency in Iraq* (Ithaca, NY: Cornell University Press, 2006), pp. 59–124.

3 Author's interviews in Iraq, November 2003–April 2004.

4 Chris Kutschera, 'The Kurds' Secret Scenarios', *Middle East Report*, no. 225, Winter 2002, pp. 14–21.

5 On the rise of political Islam among the Kurds see Michiel Leezenberg, 'Political Islam among the Kurds', in Faleh Jabbar and Hosham Dawod (eds), *The Kurds: Nationalism and Politics* (London: Saqi Books, 2006), pp. 203–30.

6 Jaza Muhammad, 'Tawhid and Jihad and al-Qa'idah's Kurdistan Brigades Intend to Move North to Kurdistan', *Awena* (Sorani Kurdish newspaper, Suleimaniya), Open Source Center, GMP20070412950030, 10 April 2007.

7 The size of the Turkmen population is a very controversial issue in Iraq. Turkey and the Turkmen in Iraq contend that the former regime downplayed the number of Turkmen in Iraq as part of its policy of 'Arabising' the Turkmen. They also contend that this policy has been continued by the post-Saddam regime, largely due to the rise of Kurdish influence. The Kurds want to play down the number of Turkmen in northern Iraq, particularly in the oil-rich region of Kirkuk and its environs and in the strategic province of Nineva.

8 Necdet Sivasli, 'Turkomans Say We Have Been Abandoned to Our Fate', *Ortadogu*, 10 October 2005, in FBIS, GMP20051010022004, 10 October 2005, https://www.fbis.gov/portal/server.pt/gateway/PTARGS_0_1696_246_203_0_43/http.

9 Author's observations in Tel Afar, Mosul and Kirkuk, August–October 2005.

10 'Growing Rifts Within Insurgency in Iraq, Part II: Baathists vs. the Army of Mujahideen', *Global Issues Report*, 11 April 2007.

11 *Ibid.*

12 *Ibid.*

13 *Al-Hayat*, 11 April 2007, Open Source Center, GMP20070411825007.

14 'Jihadist Websites: OSC Summary', Open Source Center, FEA20070504131508, 3 May 2007.

15 Jesse Nunes, 'Widening Schism in Iraq Between Sunni Insurgents, Al Qaeda', *Christian Science Monitor*, 27 March 2007, http://www.csmonitor.com/2007/0327/p99s01-duts.html.

16 Maher al-Jasem, 'Sunni Face New Conflicts in Iraq War', *Aljazeera.net*, 24 November 2006.

17 Author's observations and interviews in Baghdad at the Republican Palace ('Green Zone'), Camp Victory and Camp Slayer, October–November 2007.

18 Ned Parker, 'Insurgents Report a Split with Al Qaeda in Iraq', *Los Angeles Times*, 27 March 2007, p. 1.

19 *Ibid.*

20 Mahmud al-Zubaydi interview on al-Arabiyah Television, 9 April 2007, Open Source Center, GMP20070409641001.

21 *Ibid.*

22 *Ibid.*

23 'Prominent Iraqi Jihadist Group Denounces Al-Qa'idah in Iraq', BBC, 6 April 2007, Open Source Center, GMP 20070406950023, http://www.opensource.gov/portal/server.pt/gateway/PTARGS_0_0_200_240_1019_43/h.

24 *Ibid.*

25 *Ibid.*

26 Sudarsan Raghavan, 'Sunni Factions Split With Al-Qaeda Group', *Washington Post*, 13 April 2007.

27 *Ibid.*

28 'Mujahideen Shura Council Fires Back at Critics', *Global Issues Report*, 19 October 2006.

29 'Prominent Jihadist Says Situation in Iraq has Deteriorated since Establishment of ISI', 'Jihadist Websites:

OSC Summary', Open Source Center, GMP20070307376001, 1 March 2007.

30 'Islamic State of Iraq Lashes Out at Its Enemies and Rivals', *Global Issues Report*, 15 March 2007.

31 'The Islamic State of Iraq and Its Critics, Part I: Why We Fight', *Global Issues Report*, 13 April 2007.

32 John Burns, 'Iraqi Tribal Leader Is Killed, and Mourners Are Attacked', *New York Times*, 25 May 2007, p. 6; John Ward Anderson, 'Iraqi Tribes Strike Back at Insurgents', *Washington Post*, 7 March 2006, p. A12.

33 Khalid al-Ansary and Ali Adeeb, 'Most Tribes in Anbar Agree To Unite Against Insurgents', *New York Times*, 18 September 2006, http://ebird.afis.mil/ebfiles/e20060918456753.html.

34 'Al-Qaida: en perte de vitesse en Irak', Recherches-sur-le-terrorisme.com, January 2007, http://www.recherches-sur-le-terrorisme.com/Documentsterrorisme/qaida-irak-anbar.html.

35 Al-Ansary and Adeeb, 'Most Tribes in Anbar Agree To Unite Against Insurgents'.

36 'MSC and Al-Qaida in Iraq Denounce Anbar Tribal Leaders', *Global Issues Report*, 3 October 2006.

37 Charles Levinson, 'Sunni Tribes Turn Against Jihadis', *Christian Science Monitor*, 6 February 2006, http://www.christiansciencemonitor.com/2006/0206/p01s01-woiq.htm.

38 'Iraq's Sunni Arabs Confront the Islamic State of Iraq, Part II: Al Qa'ida Strikes Back', *Global Issues Report*, 2 May 2007.

39 See for example Seumas Milne, 'Out of the Shadows', *Guardian*, 19 July 2007.

40 'Al-Sabil Monitors the Jihadist Landscape of the Resistance Factions in the Land Between the Two Rivers', 17 July 2007, p. 19, Open Source Center, GMP20070717633003, https://www.opensource.gov/portal/server.pt/gateway/PTARGS_0_0_5762_972_100578.

41 Alan Sipress, 'Feeling Besieged, Iraq's Sunnis Unite', *Washington Post*, 6 January 2004, p. A11.

42 Author's observations and interviews, Mosul, August–September 2005.

43 Al-Jazeera TV, 11 February 2004, Open Source Center, GMP20040213000113.

44 *Al Arab al-yawm*, 23 April 2006, p. 14, Jihad al-Rantisi, 'Al-Arab al-Yawm Interviews Head of Association of Muslim Scholars in Iraq', FBIS Report, Open Source Center, GMP20060423534002.

45 Khalil Harb, 'Interview with Shaykh Harith al-Dari', *al-Safir*, 24 March 2007, p. 14, Open Source Center, GMP20070326636001.

46 Jonathan Steele, 'Iraqi Insurgents Regrouping, Says Sunni Resistance Leader', *Guardian*, 3 December 2007, http://www.guardian.co.uk/print/0,,331429878-103550,00.html.

47 'Iraq's Tribal Counterinsurgency Fuels Internal Sunni Rivalries', *Global Issues Report*, 7 December 2007.

48 Jack Fairweather, 'Political Ambitions of Sunni Tribal Leader Worry Baghdad Elite', *Financial Times*, 19 April 2008, http://ebird.afis.mil/ebfiles/e20080419594795.html.

49 Edward Wong, 'Some Insurgents Are Asking Iraq for Negotiations', *New York Times*, 27 June 2006, p. 1; 'Iraq's Tribal Counterinsurgency Fuels Internal Sunni Rivalries'.

50 Doug Smith and Saif Rasheed, 'Sects Unite to Battle Al Qaeda In Iraq', *Los Angeles Times*, 19 November 2007, p. 1.

51 'Insurgent Groups Unite to Form "Jihad and Reformation Front"', 'Jihadist Websites', Open Source Center, FEA20070504131508, 3 May 2007, https://www.opensource.gov/portal/server.pt/gateway/PTARGS_0_0_6006_989_0_43/http.

52 See for instance Jeffrey Record, *Beating Goliath: Why Insurgencies Win* (Dulles, VA: Potomac Books, 2007); Rex Brynen, *Sanctuary and Survival: The PLO in Lebanon* (Boulder, CO: Westview Press, 1990).

53 Milne, 'Out of the Shadows'.

54 *Ibid.*

55 *Ibid.*

56 *Ibid.*

57 See Abdul-Ahad, 'Outside Iraq But Deep in the Fight: A Smuggler of Insurgents Reveals Syria's Influential, Changing Role', *Washington Post*, 8 June 2005, p. A1.

58 Mounir Elkhamri, 'Iran's Contribution to the Civil War in Iraq', Occasional Paper, Jamestown Foundation, January 2007.

59 See for example 'Jihadist Website Details Methods of Setting Traps, Ambushes Against US Forces', 'Jihadist Websites: OSC Summary', Open Source Center, GMP20080304488003, 29 January 2008, https://www.opensource.gov/portal/server.pt/gateway/PTARGS_0_0_200_240_1019_43/h.

60 Daniel McGrory, 'Would-Be Martyrs Log On and Slip Across the Border', *The Times*, 22 June 2005, http://www.timesonline.co.uk/tol/news/uk/article535993.ece.

Conclusion

1 The reasons for the US decision to invade Iraq and the justifications made for the invasion have been addressed on many occasions, and there have been numerous analyses of what went wrong. There is no reason to retread these issues here. But for details on the ideological narrative of the George W. Bush administration, see Nicholas Lehman, 'After Iraq: The Plan to Remake the Middle East', *New Yorker*, 17 February 2003, http://ebird.dtic.mil/Feb2003/e20030211153280.html; Janine Zacharia, 'Building a Free and Democratic Iraq is Going to be a Huge Victory in the War on Terrorism', *Jerusalem Post*, 26 September 2003, p. 6, http://ebird.afis.osd.mil/ebfiles/e20030930220472.html. See also Johanna McGeary, '3 Flawed Assumptions', *Time*, 7 April 2003, p. 58, http://ebird.dtic.mil/Mar2003/s20030331169655.html; Rajiv Chandrasekaran, 'A Forced Retreat', *Washington Post National Edition Weekly*, 5–11 January 2004, pp. 6–7; Larry Diamond, 'What Went Wrong in Iraq', *Foreign Affairs*, September–October 2004.

2 Stephen Hedges, 'Former General Says US Military Didn't Expect Iraqi Insurgency', *Chicago Tribune*, 15 July 2004.

3 William Booth and Daniel Williams, 'US Soldiers Face Persistent Resistance', *Washington Post*, 10 June 2003, p. 1; Rowan Scarborough, 'Organized Iraq Rebels Coordinate Strikes', *Washington Times*, 17 November 2003, p. 1. For more details of the evolution of the insurgency during 2003, see Michael Knights and Jeffrey White, 'Iraqi Resistance Proves Resilient', *Jane's Intelligence Review*, November 2003, pp. 20–24.

4 Volkhard Windfuhr and Bernhard Zand, 'Religious Strife Is Pushing Iraq Towards Civil War', *Der Spiegel Online*, 6 March 2006, http://www.spiegel.de/international/spiegel/0,1518,404503,00.html; Abdul-Ahad, 'Inside Iraq's Hidden War'; Scott Peterson, 'Iraq's Deepening Religious Fissures', *Christian Science Monitor*, 28 November 2006, p. 1; Dan Murphy, 'Growing Friction Separates Shiite, Sunni', *Christian Science Monitor*, 2 March 2006, http://www.csmonitor.com/2006/0302/p07s02-woiq.htm.

5 Hashim, 'Iraq's Civil War', *Current History*, January 2007; Nir Rosen, 'Anatomy of a Civil War', *Boston Review*, 20 November 2008; Ken Silverstein, 'The Minister of Civil War: Bayan Jabr, Paul Bremer and the Rise of the Iraqi Death Squads', *Harper's Magazine*, August 2006, pp. 67–73.

6 Louise Roug and Richard Boudreaux, 'Deadly Rift Grows Among Insurgents', *Los Angeles Times*, 29 January 2006, p. 1; Charles Levinson, 'Sunni Tribes Turn Against Jihadis', *Christian Science Monitor*, 6 February 2006, p. 1; Partlow, 'Rival

Sunnis in Deadly Gunfight', *Washington Post*, 11 November 2007, p. 26.

7 Ned Parker, 'Bloody Rivalry Between Iraqi Insurgents', *Los Angeles Times*, 11 November 2007, http://ebird.afis.mil/ebfiles/e20071111560471.html.

8 Sabrina Tavernise and Dexter Filkins, 'Local Insurgents Tell of Clashes with Al Qaeda's Forces in Iraq', *Der Spiegel Online*, 12 January 2006, http://www.spiegel.de/international/0,1518,druck-394828,00.html.

9 Iraqi society should not be seen only in terms of tribes on the one hand and the cities or government on the other, or just in terms of ethno-sectarian communities; class has also played an important role in politics. For an extensive analysis, see the definitive work by Hanna Batatu, *Old Social Classes and the Revolutionary Movements of Iraq* (Princeton, NJ: Princeton University Press, 1978).

10 For a similar evaluation, see Raghavan, 'Sunni Factions Split With Al-Qaeda Group', p. 1.

11 See Steven Simon, 'The Price of the Surge: How US Strategy is Hastening Iraq's Demise', *Foreign Affairs*, vol. 87, no. 3, May–June 2008, pp. 57–76; Colin Kahl, Michele Flournoy and Shawn Brimley, *Shaping the Iraq Inheritance*, Center for a New American Security, June 2008, p. 3. In *The War Within* (New York: Simon & Schuster, 2008), veteran journalist Bob Woodward adds another reason for the decline in AQM's potency and thus the diminution in violence, namely the strategy launched by US intelligence and Special Operations Command (SOCOM) against commanders of AQM and other groups. For a summary of Woodward's argument, see 'Why Did Violence Plummet? It Wasn't Just the Surge', *Washington Post*, 8 September 2008, p. A9. Journalist Hazim al-Amin of *Al-Hayat* observed in early September 2008 that a weakened AQM, demoralised by its battering in Iraq at the hands of Sunni insurgents and US forces, had decided to return to Afghanistan, where the insurgency by the Taliban against NATO forces seems to be enjoying a new lease of life.

12 Raghavan, 'No Relief From Fear', *Washington Post*, 5 September 2007, p. 1.

13 'Iraq's "Awakening" Against al-Qaida Spreads to Baghdad – Where Insurgents Strike Back', *Global Issues Report*, 24 August 2007.

14 Author's observations, Amiriya, October–November 2007.

15 For analyses of the profound hostility and contempt that each side feels for the other, see the excellent studies by Rend Rahim Francke, 'Political Progress in Iraq During the Surge', Special Report no. 196, United States Institute of Peace, December 2007, pp. 5–10; Kanan Makiya, 'Is Iraq Viable?', Middle East Brief no. 30, Crown Center for Middle East Studies, Brandeis University, September 2008.

16 Though the Kurds have begun to be affected by al-Maliki's new-found confidence: in September 2008, the prime minister sent Iraqi Army troops to challenge Kurdish encroachment into territory outside the Kurds' autonomous region.

ΓIISS ADELPHI PAPERS

RECENT **ADELPHI PAPERS**:

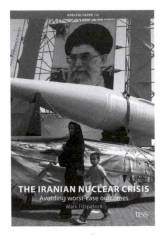

ADELPHI PAPER 398

The Iranian Nuclear Crisis:
Avoiding worst-case outcomes

Mark Fitzpatrick

ISBN 978-0-415-46654-7

ADELPHI PAPER 399

Joining al-Qaeda: Jihadist
Recruitment in Europe

Peter R. Neumann

ISBN 978-0-415-54731-4

All Adelphi Papers are £9.99 / $19.99

For credit card orders call **+44 (0) 1264 343 071**
or e-mail **book.orders@tandf.co.uk**
Orders can also be placed at **www.iiss.org**

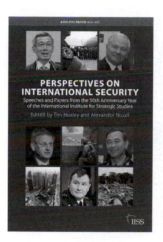

The Evolution of Strategic Thought
Classic Adelphi Papers

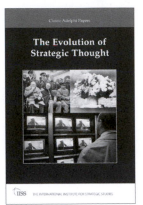

The Adelphi Papers monograph series is the Institute's principal contribution to policy-relevant, original academic research. Collected on the occasion of the Institute's 50th anniversary, the twelve Adelphi Papers in this volume represent some of the finest examples of writing on strategic issues. They offer insights into the changing security landscape of the past half-century and glimpses of some of the most significant security events and trends of our times, from the Cold War nuclear arms race, through the oil crisis of 1973, to the contemporary challenge of asymmetric war in Iraq and Afghanistan.

Published April 2008; 704 pp.

Bookpoint Ltd. 130 Milton Park, Abingdon, Oxon OX14 4SB, UK
Tel: +44 (0)1235 400524, Fax: +44 (0)1235 400525
Customer orders: book.orders@tandf.co.uk
Bookshops, wholesalers and agents:
Email (UK): uktrade@tandf.co.uk,
email (international): international@tandf.co.uk

Routledge
Taylor & Francis Group

IISS THE INTERNATIONAL INSTITUTE FOR STRATEGIC STUDIES